On the political economy of plant disease epidemics

This publication was financially supported by the
'Foundation Willie Commelin Scholten for Phytopathology'
(Stichting Willie Commelin Scholten voor de fytopathologie).

On the political economy of plant disease epidemics

Capita selecta in historical epidemiology

J.C. Zadoks

Wageningen Academic
P u b l i s h e r s

ISBN 978-90-8686-086-9

Photo cover:
potato field with beginning late blight
focus (right side, under the middle),
set in a virtual epidemic explosion

First published, 2008

Wageningen Academic Publishers
The Netherlands, 2008

To my wife, children and grandchildren who gave me the pleasure of being.

To the memory of Jan de Vries and Jo de Vries-Bosscha, and others, who gave me more than only physical shelter in a time of fury (1940-1945).

'Agriculture and gardening, though of such great utility in producing the nutriment of mankind, continue to be only Arts, consisting of numerous detached facts and vague opinions, without a true theory to connect them, and to appreciate their analogy; at a time when many parts of knowledge of much inferior consequence have been nicely arranged, and digested into Sciences.'

Erasmus Darwin – 1800. Phytologia; or the philosophy of agriculture and gardening. London, Johnson, page VII.

Preface

Plant disease epidemics can strongly affect the life and happiness of individual farmers, their dependents, and their nations. The linkage between agricultural production and the welfare of the nation and state is covered by the 18th century term 'political economy'. The branch of epidemiology that reconstructs, describes, and explains plant disease epidemics of the past I call 'historical epidemiology'. In the case studies, presented here as *capita selecta*, I consciously stepped over the demarcation line between historical epidemiology and agricultural history. The emphasis in these 'selected chapters' is not so much on the biological causes, for which I refer to the current text books, but rather on the human consequences of plant disease epidemics. Not pretending to be a historian, I followed my own fancy in looking at some of the consequences of selected plant disease epidemics on society at large.

Chapter 1. In ancient times the helplessness of people that had to face plant disease epidemics without any means of control forced them to cope with plant disease and crop loss in ways unfamiliar to the present Western mind. In classical times a severe rust epidemic on wheat, the common staple in the Mediterranean area, meant hunger to the peasants and their families, and maybe also ruin to their state. Having no knowledge of the underlying biology, they had to deal with such catastrophes in symbolic terms, by means of folk tales and religious ceremonies.

Chapter 2. Whereas in the second half of the 20th century yellow stripe rust on wheat was a familiar disease, causing occasional epidemics in all five continents, yellow rust on rye disappeared in the first half of the century. Before, however, the disease could be locally severe. The extent and impact of the yellow rust epidemic on rye in 1846 remains unmatched up to this day. The impact of this epidemic on Belgium, the Netherlands and part of the Rhine Province in Germany is demonstrated but its importance in a belt stretching from Denmark through western Germany down into Switzerland remains uncertain. The epidemic also affected the western part of modern Poland from the Carpathian Mountains to the Baltic Sea, but to what degree? The losses came at the worst possible moment, when Continental Europe was in the grip of potato late blight.

Chapter 3. Generally speaking plant disease epidemics had fewer and smaller effects than some natural phenomena such as volcanic eruptions and prolonged droughts, or historical phenomena such as political strife and war, but some epidemics had an exceptional impact on human history. The outbreak of *Phytophthora infestans* in Europe, 1844-46, stands out as a model with its unequalled social, economic, even political consequences. Historians focused on the misery brought about by the 'potato murrain' in Ireland, whereas the equally adverse but politically very different effects of the potato late blight on the European Continent have been neglected. Chapter 3 intends to remedy this deficiency by introducing the 'Continental Famine'.

Chapter 4. During World War I the Netherlands, remaining 'neutral' during the turmoil, suffered as the economy came to a stand-still. The food situation became precarious, to put it mildly. No data exist on the losses of the major starch crops due to pathogens at a time when, amazingly, the Dutch phytopathologists of the period showed little interest in food production. Applying an unconventional method to a conventional long-term data set I made some tentative crop loss estimates for wheat , rye, and potato. The results may be seen as a background for the events of the next chapter.

Chapter 5. Phytopathologists may be tempted to associate plant disease epidemics with big historical events. One instance is the role of potato late blight in the defeat of Germany during the First World War, 1914 - 1918, as indicated by two well-known phytopathologists. They may have been over-eager in their historical judgement. Life is far too complex to accept a one-to-one relationship between a plant disease and a great historical event. Political decisions and the outcome of the fighting affected the epidemics as much as the epidemics affected the decisions and the fighting. And never forget the weather!

Chapter 6. A historian associated a plant disease, ergot on rye caused by *Claviceps purpurea*, with a major historical event, the French Revolution of 1789. This revolution changed the course of European history. Though the association may not be unjustified the causation of that revolution was complex as indicated by tens if not hundreds of books on the subject. These books, however, overlooked the possible effect of ergot. Ergot of rye also caused a 'non-event' of historical importance. A military campaign that could have changed the history of Eastern Europe and the Near East did *not* happen, a forgotten paragraph in the history of plant disease. Two epidemics have been associated with famines, black stem rust of wheat with the Russian famine of 1932-33 and brown spot of rice with the Bengal Famine of 1943. Though professional historians do not deny the presence of rust and brown spot, respectively, they tend to believe that these plant diseases were not decisive in causing the famines of the Soviet Union, 1932-33, and of Bengal, 1943.

Chapter 7, contemplating the aspect of Good Governance, looks back to the foregoing chapters. It strikes a more personal note.

Jan C. Zadoks

Table of contents

List of abbreviations

[ ]	= Comments inserted by author
~	= approximately

BCE	= Before Christian Era
BP	= Before Present
CE	= Christian Era
D	= Dutch
E	= East
F	= French
G	= German
M	= middle
N	= north
NL	= the Netherlands
R	= Russian
RH	= relative humidity
S	= south
W	= west

d	= day
hl	= hectoliter = 100 liter
kg	= kilogram
l	= liter
tonne	= 1000 kg
y	= year

List of tables

List of figures

List of boxes

1. Wheat rust in antiquity

In antiquity wheat rust was known and feared. The visual effects of the rust were described in metaphorical language as red foxes and wheat fires. In classical Rome the rust was personified as a *numen* and magic was applied to propitiate it with sacrifices, processions, and religious feasts, the *Cerealia* and *Robigalia*. Which rust species was involved? The field symptoms point to black stem rust. Recent crop and rust phenology data and astronomical comments by Roman authors were in fair agreement. To describe the damage caused by rust a story was told in figurative language. Foxes, burning torches bound to their tails, were said to set fire to the wheat fields. In Rome Ovid versified the story 2000 years ago, the biblical tale of Samson is about 2700 years old, and the fable told by the legendary Aesop from Asia Minor may go back beyond 2600 before present. The three narrations, sharing the element of blazing foxes set free at harvest time, are seen as three versions of a folktale circulating in the Mediterranean area, meant as a metaphor to describe and explain the effect of a severe attack by black stem rust on wheat.

Rubigo quidem, maxima segetum pestis, lauri ramis in arvo defixis transit in ea folia ex arvis.
As for the greatest curse of corn, mildew, fixing branches of laurel in the ground makes it pass out of the fields into their foliage.

Pliny the Elder (23/24-79 CE) – *Naturalis historia* XVIII: 161.
Translation by H. Rackam (1971).

'The ancients having no rational Principles or Theory of *Agriculture*, plac'd their chief Confidence in Magical Charms and Enchantments; which he, who has the Curiosity and Patience to read, may find in the Title aforemention'd, in *Cato*, in *Varro* (and even *Columella* is as fulsome as any of them) all written in very fine Language;'

Tull, J. (1733: 68) – The horse-hoing husbandry.

1.1. Famines feared

Famines have been a matter of constant fear among the farmers of the past. The causes of famines were diverse, abiotic, biotic, or man-made. Man-made famines have been used as an instrument of warfare[1]. Abiotic famines were climate-induced and due to either frosts, droughts or floods. Among the biotic causes were pests[2] and diseases[3] of crops. In antiquity, one of the diseases felt to be a real danger and a possible cause of famine was the wheat rust.

The classical agronomists could not identify the rust species as it is done today. This chapter[4] intends to demonstrate that from ~3000 BP (Before Present) until recent times the black stem rust, *Puccinia graminis* f.sp. *tritici*, was felt to be a permanent threat to the farming communities of the Mediterranean Area and the Near East.

In olden times, the link between the natural (drought, locust, 'mildew') and the super-natural was evident to observing people. By lack of logical, not to say bio-logical, explanations of catastrophic events anxious persons, threatened in their very existence, were driven to super-natural explanations and magic[5] within the framework of their religion. Priests of all denominations drew moral lessons from agricultural mishaps and threatened their flocks to keep them on the right religious track. At home, around the fire place, people told stories which contained elements of admonition, explanation and consolation. The wording was metaphorical, even poetical, and in no way scientific in the modern sense.

1.2. Wheat and wheat rust

1.2.1. Wheat

Wheat cultivation began some 12,000 years BP, when wheat was selected as a nutritious, cultivatable and harvestable grass (Diamond, 1999). The ears of grasses tend to be fragmented for easy dispersal of the seeds but an intact ear is crucial for harvesting a grain crop. The intact ear is guaranteed by one double recessive gene. Finding this gene and maintaining it in the double recessive state was the first step in wheat[6] selection and breeding taken ~10,000 years ago.

1.2.2. Rust

Wheat can be infected by three rust species, yellow stripe rust, brown leaf rust, and black stem rust, all three now occurring world wide. These parasitic fungi have co-evolved with their host grasses[7]. The various species of grass rusts were not so choicy with regard to their host species, in the Near East up to this very day (Wahl *et al.*, 1984).

The life cycle of some of these rust species is quite complex. Several rusts are macrocyclic and heteroecious[8]. Macrocyclic means that these rusts have different spore forms with different functions. Heteroecious means that some of these spore forms, belonging to one and the same rust species, need different host plants. Thus the black stem rust has two spore forms on the common barberry (*Berberis vulgaris*), and three spore forms on grass, i.c. wheat. This polymorphism can be seen as an adaptation to the climate. As long as the grasses are lush and green, the rust multiplies and spreads by means of urediniospores. With the advent of summer the grasses begin to dry and the rust produces telia with teliospores, a resting stage that bears with adverse conditions, cold, heat and drought. When the rains set in the teliospores will germinate and produce basidiospores that can drift about in the wind. The life cycle continues when the basidiospores settle on the young leaflets of the barberry, freshly appearing after the first rains. The basidiospores infecting the barberry leaflets produce pycnia and these lead, after a fertilisation process, to aecia with aeciospores. Aecidiospores are shed and dispersed by wind to grasses, among which wheat (Lehmann *et al.*, 1937). In the space-time universe we need barberry shrubs not too far from the grass or wheat fields, and an alarm-clock awakening simultaneously the teliospores to produce infecting basidiospores and the dry barberry bushes to produce infectible young leaflets. The first rains set the alarm.

Specialisation of black rust on selected grasses occurred in the more peripheral areas of cultivation[9]. In Western Europe specialised forms of the rust exist on wheat, oats, rye and barley, on perennial ryegrass, and so on. They all alternate with barberry, but they can do without host alternation when offered a continuous, year-round crop[10]. The final choice depends on various circumstances not relevant here. Curiously, no evidence has been found in modern Israel for black rust on barberry (Anikster & Wahl, 1979).

Barberry, a thorny shrub, probably was common in the Mediterranean area. Supposedly it occurred at the fringes where grassland and forest met. In recent times four vicariant species have been distinguished which occur in a botanical association, the Berberidetum, in the mountains at mid-height[11]. Barberry is an aggressive coloniser of disturbed land and, once established, difficult to get rid of.

In Kanaänite antiquity numerous but relatively isolated villages, separated by stretches of forest and grazing grounds, were surrounded by small fields. For safety reasons villages were usually located on hill tops. The position of barberry in a landscape dominated by wilderness is not clear. It may have occurred at the fringes where grassland and forest met. Much later, the Romans established large estates in parts of the Mediterranean area, as in Tunisia, possibly with barberry bushes in the roughages left over between fields.

In modern times *B. vulgaris* has been planted as an ornamental and medicinal plant. Hedges of barberry have been planted to keep the cattle or the 'naughty boys' out[12]. Because of the black rust the common barberry had to be eradicated.

1.2.3. Evidence for black stem rust

The Bible contains several curses that threaten crops to be 'smitten with mildew'[13], and 'mildew' is an old English term for what is now called rust, primarily black stem rust, more specifically the telial stage[14]. Löw (1928: 40) related the words ירקון and שדפון to wheat diseases such as smut and black rust. Lehmann *et al.* (1937) stated that Hebrew texts juxtapose the words *yeraqon* and *shidafon*, referring to bleaching or yellowing and scorching or blackening, respectively. The bleak and the black are the final colours of a wheat crop severely infected by stem rust.

If the Biblical texts refer indeed to rust, the choice of a species becomes easy. We have to look at descriptions of 'field symptoms' from the pre-scientific period. A severe epidemic of black rust in its terminal stage may end up in a miserable, ashen-grey crop[15], sprinkled with dull black spots. The crop is 'of a dark livid colour' and looks like 'covered with soot'[16]. The rusted crop seems to be scorched, even burnt. The association of black rust with fire comes naturally. In serious cases the stems break and fall hither and thither[17], the grain is shrivelled, and the harvest is a complete failure. Wheat crops severely affected by yellow or brown rust may lead to complete harvest failure[18], though rarely so, and they may look miserable, but they do not typically remind one of fire.

The classical Greek authors were extremely vague on plant disease and so we will disregard them, one example excepted. Aristotle[19] hinted at something that could be interpreted as rust though the reference is not convincing: 'Now red blight is due to moisture which is hot, but comes from elsewhere'. The word translated as 'red blight' is *erusibai, erysiphe* – mildew – the old English term for the final stage of black stem rust[20].

Of far later date than our biblical informants are the various Roman authors[21] knowledgeable in agriculture and providing relevant information. Ovid (see Box 1.1) was pertinent as to 'rust', the

---

Box 1.1. Ovid on the Robigalia in Fasti (IV: 905 ff; translation by Frazer, 1989).
The word Robigo was retained instead of Frazer's 'mildew'. Note the pun of rust on crops and on iron.

On that day, as I was returning from Nomentum to Rome, a white-robed crowd blocked the middle of the road. A flamen [= priest] was on his way to the grove of ancient Robigo to throw the entrails of a dog and the entrails of a sheep into the flames. Straightaway I went up to him to inform myself of the rite. Thy flamen, O Quirinus, pronounced these words:

> 'Thou scaly Robigo, spare the sprouting corn, and let the smooth top quiver on the surface of the ground. O let the crops, nursed by the heaven's propitious stars, grow till they are ripe for the sickle. No feeble power is thine: the corn on which thou hast set thy mark, the sad husbandman gives up for lost. Nor winds, nor showers, nor glistening frost, that nips the sallow corn, harm it so much as when the sun warms the wet stalks; then, dread goddess, is the hour to wreak thy wrath. O spare, I pray, and take thy scabby hands from off the harvest! Harm not the tilth; 'tis enough that thou hast the power to harm. Grip not the tender crops, but rather grip the hard iron. Forestall the destroyer. Better that thou shouldst gnaw at swords and baneful weapons. There is no need of them: the world is at peace ... but let rust defile the arms, and when one essays to draw the sword from the scabbard, let him feel it stick from long disuse. But do not thou profane the corn, and ever may the husbandman be able to pay his vows to thee in thine absence.'

So he spoke. On his right hand hung a napkin with a loose nap, and he had a bowl of wine and a casket of incense. The incense, and wine, and sheep's guts, and the foul entrails of a filthy dog, he put upon the hearth – we saw him do it. Then to me he said,

> 'Thou askest why an unwonted victim is assigned to these rites?' Indeed, I had asked the question. 'Learn the cause,' the flamen said. 'There is a Dog (they call it the Icarian Dog), and when that constellation rises the earth is parched and dry, and the crop ripens too soon. This dog is put on the altar instead of the starry dog, and the only reason of killing him is his name.'

---

same that defiled the iron arms, with the rust-brown colour[22] of the urediniospores. The field symptoms, as perceived by the Romans, and the late appearance of the disease, after the barley harvest[23], both point to black rust, *maxima segetum pestis*[24] (the worst pest of the crops). Pliny the Elder (23/24-79 CE) characterised the weather conditions[25] inducive to a rust outbreak, clear night skies and warm days.

The biological evidence for black stem rust on wheat is quite convincing. Kislev (1982) found black stem rust on charred remnants[26] of wheat (*Triticum parvicoccum*) in an Israeli dig dated ~3000 BP.

1.3. Magic

1.3.1. Magic and science

Magic and science have in common that, in principle, a one-to-one relationship exists between an action, the 'cause', and its result, the 'effect'. In science the relationship is mechanistic, in magic it is associative. The separation line between magic and religion is thin. Magic implies that a specific desirable result follows from a specific human action[27]. Sacrifices and processions, religious feasts and games, so common in antiquity all contain elements of magic.

In this chapter the Roman feasts called 'Cerealia', the grain feast, and 'Robigalia', the rust feast, will be highlighted.

1.3.2. Cerealia

The Cerealia (April 12 to 19) were devoted to the cereals[28]. This feast contained elements of magic. Ovid (43 BCE-17 CE, Fasti IV: 679 ff) described a curious event during the Cerealia. Foxes, with burning torches bound to their tails[29], were chased through the *circus maximus* in Rome[30]. As an 'explanation' Ovid told a little story. A naughty boy, a farmer's son, wanted to punish a fox that had stolen many farmyard fowls by wrapping it 'in straw and hay, and setting a light to her'. The fox escaped. ´Where she fled, she set fire to the crops that clothed the fields, and a breeze fanned the devouring flames'[31]. The breeze spreading the flames over the fields suggests the contagious nature of the rust disease that, carried by the wind, affects an area rather than a single field. Apparently, the 'wheat fire' was noticed just before harvest, when the wheat was dry and burned easily. This timing is relevant, as we will see.

1.3.3. Robigalia

The Robigalia are known from Ovid (see Box 1.1). Roman calendars date the feast at April 25th, nearly the date Ovid quoted for his encounter with a festive procession led by a sacrificer (*flamen*). The Praenestine calendar[32] added 'The festival of Robigus takes place at the fifth milestone on the Via Claudia, lest Robigo should harm the corn. A sacrifice is offered and games are held by runners both men and boys'.

Robigo (female, threatening) or Robigus (male, protecting) originally was a *numen*, an abstract notion, neither male nor female, representing a phenomenon (Leopold, 1926), e.g. a spring or a special tree. *Numina* were commonly worshipped by the Romans. To appease the *numen* in good time a ceremony was organised, the Robigalia. Pliny stated that this feast was established by the Roman King Numa Pompilius (715-672 BCE) at ~700 BCE. Possibly, the king formalised an older, possibly Etruscan, tradition.

Indeed, the ceremony was established at an early date as follows from the location, at the fifth mile stone from the centre of Rome, once the border of the Roman *ager*, the city's cultivated area. It is at the border where one has to defend his crop. In Ovid's time the Robigalia were apparently a folkloristic event for townspeople rather than a religious ceremony for peasants.

Passing to more recent times the Christian Fathers Tertullian, Lactantius and Augustine referred to Robigo. Emperor Constantine made Christianity the State Religion in 323 CE, keeping the Robigalia in the calendar. Pope Gregory I (540-604) replaced the Robigalia in 598 by the *Festum S. Marci Evangelista*, still on April 25th. The christianised procession should never be postponed, according to a 1756 missal[33]. It followed the old route northward along the Via Flaviana (the present name of the first part of the Via Claudia), passed the Ponte Milvio, and then returned to St Peter's cathedral to celebrate mass.

April 25th and the three days preceding Ascension Thursday used to be called the Rogation Days. Processions went out into the fields, invoking the benediction of the crops by singing the litanies (see Box 1.2). Until the mid 20th century the Rogation Days were observed in many parts of W Europe including the Netherlands (Goossen, 1980; various personal communications[34]).

Box 1.2. The litanies of the Rogation Days.

On April 25th, the Major Rogation Day, the *Litaniae Majores* (major litanies) were sung. They contained vocations to Saints, prayers and supplications (*rogationes*). A major rogation is:
 Ut fructus terrae dare et conservare digneris.
 (That You deign to give and preserve the fruits of the earth).
It was the only rogation repeated thrice (Verbruggen, 1957).
On the other rogation days, called Minor Rogation Days, the *Litaniae Minores* sufficed, a shortened version with the rogation:
 Ut fructus terrae dare et conservare digneris, te rogamus, audi nos,
 (That You deign to give and preserve the fruits of the earth, we pray You, hear us).
This rogation was not repeated.
The Litaniae Majores were sung at a fixed date (April 25th) and the Litaniae Minores at a variable date, depending on the date of Eastern. These two dating systems, fixed and variable, existed already in Roman times.

The *Litaniae Majores* are no longer sung except in Rome, and since the Second Vatican Council, 1970, the Rogation Days are no more in the *Missale Romanum*. Thus a tradition nearly three millennia old was abolished.

The Church (now we say the Roman Catholic Church), when spreading Christianity over Europe, appeased local populations between the fourth and ninth century by christianising local custom. Field processions in spring formed part of Gallic and Germanic traditions. In the Netherlands, the protestant version of the Roman Catholic field processions is a day of prayer for the crops, on the second Wednesday in March, observed up to this very day.

1.3.4. Sacrifices and processions

The Roman author Varro (116-27 BCE) wrote about 'twelve gods who are the special patrons of husbandmen'. Among them were 'Robigo and Flora[35]; for when they are propitious the rust will not harm the grain and the trees ... wherefore, in honour of Robigus has been established the solemn feast of Robigalia ...'[36]. Sacrifices to the rust god Robigus[37] were made alongside the fields with standing corn[38]. Altars used to be temporary constructs of sods (Leopold, 1926). Ancient Italian farmers carved their gods in wood. Sods and wood leave no remnants.

In 1922 a stone cylinder was found in a vineyard at Castiglioncelli, near Livorno, some 200 km north of Rome (Galli, 1924; Figure 1.1). Measuring 88 cm in height and 71 cm in diameter, and weighing some 800 kg, it consisted of reddish flint-stone quarried from nearby mountains. A semi-spherical basin was carved out at the upper side, 35 cm deep, with two drains in which traces of lead pipes were found. The decoration of the cylinder's outside consisted of three oxen heads, four dogs (two sitting and two standing with curling tails), festoons with fruits, and a horned human head with snakes around the face. The human head supposedly represents the evil goddess Robigo[39]. Horned oxen and horned human heads represent apotropaeic, evil averting forces. The altar, dating from the first century BCE, may be the only material remnant of the once lively Robigo cult.

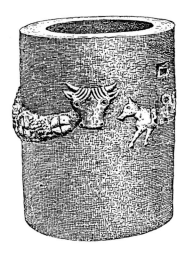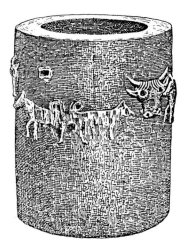

Figure 1.1. Drawings after the Robigo altar, found at Castiglioncelli (from Galli, 1924).

1.4. Science

1.4.1. A problem of timing

Roman authors provided information that can be interpreted with today's knowledge. Our interpretation so far suggested an identification of the rust involved, a choice out of three possible rust species. But there is more to it. The Roman authors were specific about dates and we may investigate whether these dates make sense to us. Can they be used to confirm or refute our interpretation? Three aspects of timing ask our attention, crop phenology, rust phenology, and astronomy. We admit that the phenological argument is 'softer' than the astronomical argument.

1.4.2. Phenology, the crop

Little is known about wheat phenology in antiquity. As the climate of Rome probably did not change much between 500 BCE and today some of the more recent agricultural data may apply. Piero de Crescentio (~1300) wrote a treatise *De agricultura vulgare* (On common agriculture). He was a citizen of Bologna, a town at the southern edge of the Po valley, 250 km N of Rome. He differentiated cold, normal and warm places. His data for warm places supposedly reflect the situation in Latium, the wheat growing area near Rome. In Book XII he states:

November	Best period for sowing wheat in warm places.
February/March	Period for weeding the cereal crops.
May	Flowering period.
June	Early June – harvest period of barley; In warm places the wheat harvest is completed towards the end of June[40].

These dates are roughly confirmed by the isophanes of wheat harvests (Azzi, 1930), the Agro-ecological Atlas (Broekhuizen, 1969), and by mid 20th century data from the International

Yellow Rust Trials (Table 1.1; Zadoks, 1961: §15). Billiard (1928) and Stevens (1942) placed the antique wheat harvest in June or early July.

One side remark has to be made. Italian wheat breeding since the 1930s aimed at early harvests in order to avoid the black rust[41]. As to the damage caused by black stem rust an advance of the harvest by one week can make a world of difference. It was known for long that late crops were more severely damaged than early crops[42], as in Tuscany, 1766 (Tozzetti, 1767).

Table 1.1. Data from rust interception trials near Rome.
Rust was observed in 14 out of 20 years (70%). For ten years adequate data were available. The mean date of rust appearance, irrespective of rust species, was 101 days (σ = 22) from January 1st, corresponding with April 11th. The calculation, based on classical epidemiological theory (Zadoks & Schein, 1979), uses several questionable assumptions. The end result is an approximation only. Original data published by kind permission of the late R.W. Stubbs.

Year	Date of observation (day/month)	$t_2{}^a$	$x_{max}{}^b$	Rust species[c]	Δt^d	$t_1{}^e$
1962	10/07	191	+	T	-	-
1963	-	-	-	-	-	-
1964	?	-	1.00	GS	69	-
1965	05/05	125	+	T	-	-
1966	26/05	146	0.05	S	31	115
1967	06/06	157	-	-	-	-
1968	04/06	156	1.00	ST	69	87
1969	04/06	155	+	T	-	-
1970	30/06	181	1.00	GST	69	112
1971	-	-	-	-	-	-
1972	15/06	167	1.00	S	69	98
1973	-	-	-	-	-	-
1974	04/06	155	0.90	GST	42	113
1975	-	-	-	-	-	-
1976	-	-	-	-	-	-
1977	04/05	124	1.00	GST	69	55
1978	02/06	153	1.00	S	69	84
1979	31/05	151	0.80	GST	53	98
1980	18/06	170	0.10	ST	35	135
1981	15/06	166	0.80	GST	33	113

[a] Date of observation, Julian day (in days from January 1st).
[b] x_{max} = maximum severity observed, espressed as a fraction ($0 \leqslant x \leqslant 1$).
[c] *Puccinia* species observed: G = *P. graminis*, S = *P. striiformis*, T = *P. triticina*).
[d] Δt = estmated time in days from rust appearance to rust observation. $\Delta t = r^{-1} *$ (logit x_2 – logit x_1) with $x_1 = 0.0001$ and $r = 0.20$; r = apparent infection rate.
[e] Estimated date (in Julian days) of first appearance of rust, $t_1 = t_2 - \Delta t$.

1.4.3. Phenology, the rust

The three wheat rusts occurred, in all probability, in the ancient fields as they do in more modern crops. In recent times (~1960-1980) rusts appeared in the area of Rome around mid April (Table 1.1). Yellow rust appeared first but remained insignificant. Brown rust appeared later, sometimes became rather serious but seldom devastating. Black rust usually came last. Though the order of appearance may have been the same in antiquity, being determined by the temperature preferences of the rusts[43], the dates of appearance and the quantities of primary inoculum may have differed in past and present.

In antiquity barberry bushes may have been more frequent than today between fields and along rivers and rivulets. If so, primary inoculum of black rust could have been more abundant than today. The scientific literature describes local or at best regional epidemics of black rust with barberry as their source[44]. Banks (1805) mentioned barberry-based local epidemics of black rust in England, and a severe and general outbreak of black rust on wheat in 1804. A regular source of local epidemics was found in Central France (Massenot, 1961). The author saw such epidemics in Bavaria (Germany) and in former Yugoslavia around 1960; they must have occurred in Italy.

Tozzetti (1767) mentioned several years with local or regional rust epidemics. He agreed with Fontana (1767) that a severe black rust epidemic raged all over Italy in 1766. The summer of 1765 was bad resulting in poor seed quality. The winter 1765/6 was cold, crops were late and thin with a poor colour. Rust arriving early wrought heavock. The relation between the appearance of rust, warm days and cool nights with dew and early morning fog was known to Tozzetti.

During the build-up phase of the epidemic the summer spores (urediniospores) are formed in great quantity. The crop may acquire a reddish hue. The farmer entering the field will find his clothes and his dog's fur coloured brown by rust spores. These sights are threatening to a man who already foresees his family going hungry during the next winter. When the wheat ripens the rust, forming telia, changes colour. The crop expected to ripen with golden yellow colour turns ash-grey and – when severely affected – grey-black. The crop looks as being scorched by fire. Stems and ears are densely covered by black telial flecks of the black rust, suggesting the charred remains of a fire (see Box 2.2).

In classical nor in mediaeval texts references were found to severe rust epidemics devastating the wheat crops. Authoritative history books[45] did not mention wheat diseases. Nonetheless rust was feared as shown by an inscription on a border stone found in Tunisia and dating from the second century CE (see Box 1.3).

Box 1.3. Invocation on a border stone found in Tunesia[46]. Among the deities are Egyptian, Semitic non-Hebrew and, maybe, local gods (Anonymous, 1996).

'Oreobazagra Oreob[azagra] Abrasax Semeseilam Stenakhta Lorsakthè Koriaukhe Adonaie, sovereign gods, hinder, turn aside from this property and from what is growing on it – in the vineyards, the olive-groves, in the seeding places   hail over the produce, grain rust, fury of the Typhonian winds, a swarm of harmful locusts, so that none of these pernicious things touch this field nor any of the produce in it; but guard them altogether unharmed and uncorrupted, as long as these stones engraved with your sacred names are here lying about the land'.

1.4.4. Astronomy, the Dogstar

Farmers in past and present tend to look at the sky when musing over a decision. The modern distinction between astronomy and meteorology did not exist in antiquity; see Pliny's Natural History, Book XVIII, which abounds with remarks on stars and weather in the typical combination of traditional farmer knowledge. Sirius, the Greater Dog Star, had a prominent position in the agrometeorology of the Mediterranean area. The star was visible most of the year. At the latitude of Rome Sirius was invisible only during ~85 days. About 700 BCE, when the Robigalia were founded, the Dog Star disappeared from the evening sky ('setting') on April 25th and reappeared in the morning sky ('rising') on July 20th (Clark, 2007). Its morning ascent announced the flooding of the Nile.

The Dog Star is in the constellation of the Dog, Κυων in Greek and *Canis* in Latin, which precedes the hunter Orion. Sirius, Σειριος in Greek (= the scorching), is one of the clearest stars of the firmament. Its Latin epitheton was *flagrans* (= the blazing). The real hot days of summer, the Dog-days, began when Sirius became visible in the morning sky (heliacal rising[47]). Pliny placed the event at 23 days after solstice (June 24th for him) or July 17th. The Dog-days[48], *dies caniculares*, were roughly between July 17th and August 8th. It was the period of the *celestis sterilitas*, the 'sterilizing influence of the heavens'[49], the hottest period of the year.

Ovid clearly stated the relation between the Robigalia and the Dogstar, Sirius. The ceremony that he described is usually seen as a sacrifice to avert damage by the summer heat during the dog days (e.g. Blaive, 1995: 281). This view seems illogical since the crops were harvested before the peak of the summer heat. Bömer (1958: 270) called the relation '*sekundär*' but I think, on the contrary, that it is crucial. The snag is in the timing. Ovid's spokesman timed the premature ripening of the crops *quo sidero moto*, when the star moves, apparently to be read as 'sets'. Thus, the true relation is with the Dogstar setting in the evening twilight[49a]. This 'heliacal setting' of Sirius occurred in Ovid's time about May 1st and in 700 BCE about April 25th. The date of the ceremony recalls its ancient origin[50].

Pliny wrote[51]: 'Numa in the eleventh year of his reign established the Feast of the Robigalia, which is now [~50 CE] observed on April 25th, because that is about the time when the crops are liable to be attacked by mildew [= rust]' and 'but the true explanation is that on one or other, according to the latitude of the observers, of the four days from the 29th day of the spring equinox [April 19th] to April 28 the Dog sets, a constellation of violent influence in itself …'. The setting of Sirius toward the end of April announced a spell of trouble and mischief. To the

Romans, it seems, the heliacal setting of Sirius could lead to evil such as rust (see also Vegoia's curse, §1.5.5).

A chain of associations comes to mind, rust, blackened crops, summer burn, Dogstar, dogs, sacrifices. There was indeed an association between rusts and dogs by way of a star, an association that made sense to the ancients. Ovid, the worldly poet, was mistaken in his timing. Pliny, the matter-of-fact scientist, was precise in his timing without going into explanations. Coincidence, year after year, suggests causality. Sober-minded Pliny rejected coercive causality[52]: 'And in this matter admiration for Nature's benevolence suggests itself, as to the fact that '..., because of the fixed courses of the stars this disaster cannot possibly happen every year, ...'.

1.4.5. Summing up

Consensus existed already about the interpretation of the Roman 'rust' on wheat as 'cereal rust' in the modern sense. My text interpretation clearly suggests black stem rust (*Puccinia graminis*) as the main candidate. An excursion into the realm of the natural sciences strengthens the suggestion, but I admit that the other two rusts of wheat, and even other foliar diseases of wheat, cannot be excluded completely. Returning to the literary trail we now aim at a wider view.

1.5. Rust, dogs and foxes

1.5.1. Associative thinking

Ovid's charming little story (§1.3.2) related a burning fox to wheat in fire, a case of associative thinking. Scorched wheat supposedly is a metaphor of the visual effect of black rust, more specifically of its telial stage. The uredinial stage precedes the telial stage, just as the fox in the story precedes the wheat fire.

1.5.2. Samson and the Philistines

A comparable story is found in the Bible (Bömer, 1956). Judge Samson is depicted as an impressive but unpleasant fellow, a womanizer, a vindictive person. He was a great foe of the Philistines. Judges 15, verses 4 and 5 tell us '4. And Samson went and caught three hundred foxes, and took firebrands, and turned tail to tail, and put a firebrand in the midst between two tails. 5. And when he had set the brands on fire, he let them go into the standing corn of the Philistines, and burnt up both the shocks, and also the standing corn, with the vineyards and olives.' (King James Bible).

One explanation of Samson's story sees the foxes[53] as the wildly moving clouds of a thunderstorm and the firebrands or torches as the thunderbolts (Bakels, 1946). However, the torrential rains during thunderstorms will rapidly extinguish fires kindled by lightning allowing at best a small area to be burnt. We might also think of dry lightning[54] kindling spontaneous wheat fires as occur in the U.S.A., but wheat fires due to dry lightning have not been observed in present-day Israel[55].

In Mediterranean antiquity, mixed cropping of 'planted' crops (olives, vines) and 'seeded' crops (wheat, pulses) was common[56] (White, 1970). Thus, if the wheat was set on fire the 'vineyards and olives' could be scorched or even burnt. Samson himself might have ignited the wheat of the Philistines. Note the timing 'in the time of wheat harvest' (Judges 15 verse 1).

A phytopathological explanation of the two verses seems attractive, an outbreak of black stem rust, though the blackened vineyards and olives stand in its way. We might think of the precipitation, rain and dew, necessary to create a severe black rust epidemic, a wetness that could also induce the growth of sooty molds on fruits, and even leaves, of tree crops giving them a dull black felty cover. Curiously, Pliny mentioned rust in wheat and scorching in vine[57] in one sentence, but that comment bears no relation to fire.

1.5.3. Aesop's fable

Ovid's lovely sketch of the farmer's son may have been inspired by one of Aesop's fables (Bömer, 1956), 'The man and the fox' (Chambry, 1960). In free translation it runs as:
> 'A man had bad feelings about a fox that caused him damage. He caught it and in revenge he attached flax imbibed with oil to its tail, and put that to fire. But a demon directed the fox to the fields of the catcher. It was harvest time. The man followed, deploring his lost harvest.'

Note the ominous 'it was harvest time', the wheat ripe and dry in the field, as, literally, in Judges 15 verse 1, and, indirectly, in Ovid's poem.

1.5.4. Dog sacrifices in antiquity

The dog-and-sheep sacrifice by Ovid's *flamen* does not stand alone. Sacrifices of dogs belong to an Indo-European tradition, often in a military context of purification or of the aversion of evil (Blaive, 1995). A dog sacrifice was mentioned in a text from ~700 BCE, engraved in bronze tablets[58]. A skeleton of a dog, probably a sacrifice, was found near a wall of Rimini, a city founded in 268 BCE; it could symbolise a watch dog (Gianferrari, 1995). The threat to be averted might have been directed either at the town itself or at the *ager*, the cultivated area around the town, producing part of the town's wheat supply. In the latter sense we may understand some references to regular dog sacrifices by Pliny and Festus (see notes 59-61).

Columella[59], referring to Etruscan religious practices, recommended to sacrifice a sucking pup. Its blood and bowels served to propitiate goddess Robigo. The colour of the pup was not mentioned. Confusion arises because of another dog sacrifice, meant to appease the scourging summer heat personified by Sirius, the Dog-Star.

Allegedly Rome had a gate called the Doggy Gate, named after the Doggy Sacrifice[60]. Little detail is known about this sacrifice, but it is certainly not the sacrifice of Ovid's *flamen*. The Doggy Sacrifice with augury was to be made around May 1st, 'before the corn has sprouted from the sheath, but not before it is in the sheath', i.e. at the booting stage[61]. In this rite of variable date (*feria conceptiva*) the dog, sacrificed to avert the ire of the Dogstar from the crops, should be red or reddish[62], the colour of both the she-fox and the uredinia of stem rust. The red could refer to the summer heat as well as to the rust, the former being the usual interpretation and the latter the author's view.

We see two story lines crossing at the dog-and-sheep sacrifice to Robigo. The first is a rather consistent line of stories on dog sacrifices for purificatory, apotropaeic or propitiatory purposes. The second is a rather accidental story line on 'fox and fire' for allegorical description and phantasmagorical explanation.

1.5.5. Linking East and West

Could it be that poorly understood natural phenomena were phrased in metaphorical terms common to the Mediterranean world in the last millennium BCE? The resemblance between the three stories, Samson's deed, Aesop's fable, and Ovid's tale, is too obvious to let it pass unnoticed.

The link between the 'foxes and fires' of Ovid and Samson may be old and precede by far the dates quoted above. Dog sacrifices were found in digs of Etruscan remnants. Three complete skeletons from the 3rd century BCE have been found near an Etruscan temple, a dog and a sheep in one pit and a fox in another pit (Gianferrari, 1995). Does the trail lead back from Ovid's stories to the Etruscans?

Indeed, the trail leads to the Etruscans, and beyond. Cheats who displaced field border stones, trespassing Etruscan law, were to be punished: *fructus saepe ledentur decutienturque imbribus atque grandine, caniculis interient, robigine occidentur* (driving rains and hail will damage and lodge the crops, when the doggy days set in, and the crops will perish by rust[63]). This is Vegoia's curse, dated ~90 BCE but of older Etruscan descent (Valvo, 1998). Ominously, the curse continues *multae dissensiones in populo* (there will be much dissent among the people). Though the Doggy Days are usually related to the summer heat following the heliacal rising of the Dogstar, they might be connected as well with its heliacal setting. In springtime the Scirocco may blow, the hot, dry and dusty wind coming straight from the Sahara Desert[64]. Even today the Scirocco causes people in Rome to be irritated and nervous. The prophesy seems to confirm the nervousness, disquietude, and unrest in that period of the year.

The Etruscans probably descended from people originating from the North-eastern part of Asia Minor (Van der Meer, 2004; Achilli *et al.*, 2007). Their predecessors may have arrived in Italy around 1200-1000 BCE. They were connected to the 'Sea Peoples', fierce pirates from the Aegean area, who ravaged the East Mediterranean coasts around 1200 BCE. According to an Egyptian source the Philistines were among them. The Sea Peoples destroyed major cities in Kanaan such as Ashdod and Ekron (Finkelstein & Silberman, 2002).

'Fox and fire' as a cultural link between the wheat fields wrecked by Samson ~2600 BP and the sacrifice to Robigo sung of by Ovid ~2000 BP is far-fetched but not impossible as there was international cultural intercourse. Philistine peasants, having brought the tale from Asia Minor, may have shared it with their Jewish neighbours.

In antiquity, international travel and exchange was common, either in trade, tourism, or warfare. A late but relevant example of a cultural link among the peoples of the Mediterranean is given by a votive stone dated ~200 CE, marking the border of a domain in present Tunisia, at the time one of the grain-sheds of Rome. The stone bears a text in Greek (see Box 1.3). In the polytheistic culture of Rome the invocation of many Gods was apparently acceptable. Among them was Adonai, the Lord of the Hebrews.

The three 'fox and fire' tales must have a common root. The Roman festivals, Robigalia and Cerealia, are said to be installed by the second king of Rome, Numa Pompilius[65] (715–673/2 BCE). They may have been of Etruscan origin or have been imported from Asia Minor by the Etruscans. The book of Judges containing Samson's story, part of the 'Deuteronomistic History', was put on record somewhere between the reign of King Josiah (639-609 BCE) and the Babylonian Exile (586--440 BCE) (Finkelstein & Silberman, 2002). The story may have been original, may have been shared between Jewish and Philistine farmers, and may have been part of the cultural luggage carried by the Philistines from Asia Minor. The body of Aesop's fables was a product of classical Greek culture, about the third century BCE. Chambry (1960) saw some consensus about a Phrygian or Lydian origin of the legendary 'Aesop', possibly in the sixth century BCE, in Asia Minor again.

Could the revenge of Ovid's farm boy and Samson's wrath represent the western and eastern branches, respectively, of a folk tale tradition[66] originating in Asia Minor and common to the populations of the East Mediterranean area?

1.6. Conclusions

1. Ovid (Fasti IV: 905 ff) referred to the heliacal setting of the Dogstar, Sirius, when the wheat rusts used to appear.
2. Ovid's 'fox and fire' was a western and Samson's 'fox and fire' was an eastern version of a common folk tale, which may have originated in Asia Minor before the 6th century BCE.
3. The 'fox' and the 'red-coloured dogs' refer symbolically to the uredinial stage of the black stem rust on wheat.
4. The 'fire' alludes to the telial stage of the black stem rust on wheat.
5. The crucial moment of the 'fox and fire' stories is 'it was harvest time', when black stem rust in its telial stage is most prominent and ominous.
6. 'Fox and fire' is a metaphor of a severe infection of wheat by black stem rust (*Puccinia graminis*), an infection that could fill the farmer with a nearly mortal fear.

2. A yellow rust epidemic on rye, 1846

The year 1846 was a year of disaster in the Netherlands, and beyond. The potato harvest failed bitterly due to potato late blight. The grain harvest also failed, for a variety of reasons. The major cereal of the Netherlands was rye. Historians mentioned the failure of the rye harvest without discussing its cause. This paper argues that yellow rust was a major cause, though few scientific papers are available to prove the point. The evidence had to be brought together by bits and pieces, taken from a variety of sources. In the end a fairly consistent picture of a yellow rust epidemic on rye could be drawn, with sudden appearance over an area reaching from the Netherlands, over Belgium, the western half of present Germany, the western part of modern Poland, and touching Switzerland. In apparent contradiction rye suffered considerable damage by drought in 1846 over large areas.

Nefaria ista pestis anni 1846...
This darned pest of the year 1846...
> From a letter by B. Auerswald, Saxony, to J.-H. Léveillé, France (1848: 777).

2.1. A year of disaster

1846 was a year of disaster in the Netherlands. Economic recession, poverty, scarcity, even hunger troubled the Dutch populace. The commoner was used to eat rye bread supplemented by potatoes or, more modern, potatoes supplemented by rye bread. Unfortunately, the potato harvest failed in 1846 due to *Phytophthora infestans* Berk., as in the preceding year. The wheat harvest also failed in 1846 after inundations here and an outbreak of voles there (Chapter 3). Rye was still the main food crop of the Netherlands and the rye harvest was a failure too because of a rust. Which one?

There is little to tell about the cultivation of rye. Winter rye was grown in all districts and on nearly all soils. Spring rye was far less important. All rye was susceptible to all three rusts, black stem rust (*Puccinia graminis* f.sp. *secalis*), brown leaf rust (*Puccinia recondita*), and yellow stripe rust (*Puccinia striiformis* f.sp. *secalis*). At the time, these rusts were still poorly known. Scientists did not yet have the vocabulary needed to accurately describe what they saw. So the evidence for one or the other rust had to be pieced together from little scraps of information accepting the risk of a mistaken conclusion.

Contemporaneous scientific literature was practically silent about the rust in rye *anno* 1846, possibly because the scientists of the day were fully occupied by the potato murrain (Chapter 3). The newspapers, in contrast, were quite communicative about the rye and its rust. One of the leading newspapers of the Netherlands, the *Nieuwe Rotterdamsche Courant*, kept its readership between hope and fear. For the month of July, the harvest month, I counted 26 scattered notes on the rye harvest in the Netherlands and 22 on rye beyond the Dutch border. This commercial intelligence, usually containing strictly local and rarely regional information, must have been of great value to the trading and shipping companies of Rotterdam.

2.2. Prelude

2.2.1. The state of the art

Around 1800, professional scientists were rare but many persons had sufficient leisure time and ambition to indulge in the natural sciences. These were largely descriptive, trying to name and order the confusing diversity offered by nature (Foucault, 1966). Experimentation, common in contemporary physics, chemistry and medicine, was still exceptional in biology. Fontana (1767: 19) had performed an 'experiment', his own word, by placing rust spores under the microscope. The period around 1800 was the epoque of general and professional encyclopaedia and handbooks. Books were published on mycology, botany and agriculture[67]. Several books mentioned diseases of cereals. Writings specifically devoted to plant disease were still rare (Fabricius, 1774; Plenck, 1794).

Desmazières (1812) had a chapter on rust[68] (F: *rouille*) in his treatise on grain production. He considered rust on rye less fearsome than on wheat. 'The Rust is a yellow dust, colour of iron rust, that one observes on the stems and leaves of a large number of plants, and particularly, from the month of April, on those of wheat. That dust there forms linear and parallel flecks'. The globules [= urediniospores] were spherical or ovoid. They were 'real intestinal plants, analogous to those of the bunt, and of the genera *Uredo* and *Puccinia*'.

The wording does not exclude yellow rust though the colour indications are somewhat contradictory. The linear and parallel flecks are a differential characteristic for yellow rust but at the time the name for black stem rust was *Uredo linearis*, described as 'with linear and parallel pustules'. The shape of the urediniospores is not distinctive either. The timing, as of April, pleads against black rust, is in favour of yellow rust but does not exclude brown rust. Desmazières, a capable mycologist, was influenced by the naturalist De Candolle (1815).

Seringe (1818) in Switzerland also described rust on cereals. He considered the rust to be a parasitic fungus. He chose his adjectives carefully; a rust on sedges was reddish from the beginning, brown when old, and its pustules were sparse. Seringe's words[69] (quoting De Candolle) 'oval pustules … extraordinary small, but normally very numerous' and 'yellow dust' recall yellow rust. His description of the urediniospores excludes black rust. The damage could be severe. In his herbarium Seringe deposited a sample taken from rye that had not even headed, probably due to 'exhaustion that the rust had produced'. The severe early infection and the numerous but small pustules suggest yellow rather than brown rust.

The state of the art was such that the three rust species on wheat and rye had not yet been clearly separated (see below). Rather, the line was drawn between the uredinial and telial stages, as by Fontana (1767), Persoon (1801) and Banks (1805). Cohabitation of, in today's terminology, urediniospores and teliospores in one pustule was known but not really appreciated[70].

2.2.2. Three cereal rusts

The taxonomist rather than his taxonomy is relevant to the present subject since the taxonomist often made a side-remark of interest to the epidemiologist. Unfortunately, the nomenclature of cereal rusts has changed so much in the course of time that back-tracking is a bit tricky. For today's nomenclature I follow Savile (1984). As European and American vernacular names differ, Europe's black rust being America's stem rust, I begin with combined names and continue with the European designations (Box 2.1)[71].

Box 2.1. *The three rusts of rye.*
Collated and simplified (after Rapilly, 1979; Roelfs, 1985; Savile, 1984; Zadoks, 1961).

European name:	**Black rust**
American name:	Stem rust
Latin name:	*Puccinia graminis* (Persoon) f.sp. *secalis* (E. & H.)
Alternate host:	*Berberis vulgaris* L. and other barberry species
Plant parts affected:	Leaves, leaf sheaths, stems, and ears (glumes sporulating mainly on the outside).
Ecology:	Prefers higher day temperatures (optimum ~18 °C), cool dewy nights.
Spring sources:	Local barberry bushes, sometimes long distance dissemination.
Oversummering:	As telia on straw, also as uredinia on self-sown plants.
Overwintering:	As telia on straw, in warmer climates as uredinial mycelium in fall-sown crops or volunteer plants, and possibly on some grasses.

European name:	**Brown rust**
American name:	Leaf rust
Latin name:	*Puccinia recondita* (Roberge *ex* Desmazières, 1857)
Alternate host:	*Anchusa* and *Echium* spp.; some other *Boraginaceae*.
Plant parts affected:	Leaves.
Ecology:	Temperature preference intermediate, sporulates during mild winters.
Spring sources:	Uredinia overwintering on rye, aecia on *Anchusa* and *Echium* spp.
Oversummering:	As telia on straw, also as uredinia on self-sown plants.
Overwintering:	As telia on straw, as uredinia on volunteer plants and fall-sown crops.

European name:	**Yellow rust**
American name:	Stripe rust
Latin name:	*Puccinia striiformis* (Westendorp) f.sp. *secalis* (E. & H.)
Alternate host:	Not known, probably non-existent.
Plant parts affected:	Leaves, leaf sheaths (sporulating at the ad-axial side only, rare), stems (non-sporulating, very rare), and ears (common, glumes mainly sporulating at the inside).
Ecology:	F.sp. *secalis* - no data; f.sp. *tritici* - adapted to low temperatures, sporulating and multiplying during mild winters, optimum temperature 7/12 °C, stops growing and sporulating at day temperatures >20 °C.
Spring sources:	Fall-sown crops, volunteer plants.
Oversummering:	As uredinia on self-sown plants and on very late crops.
Overwintering:	As uredinial mycelium in fall-sown crops and volunteer plants, also under snow cover, as long as host plants survive.

Black stem rust. Black rust of rye became very important in Denmark following the 'division of the commons'. Barberry hedges had been planted around 1800 to mark the divides between fields and very severe epidemics of black rust occurred soon after (Hermannsen, 1968). Outbreaks of black rust on rye following the planting of barberries was also seen in Germany around 1804 (Windt, 1806).

In England black rust was locally important and in some years (1804) it caused severe damage on wheat (Banks, 1805; see Box 2.2). We have no reason yet to believe that long-distance dispersal of urediniospores by winds was important at the time; nonetheless, it may have occurred (Hogg *et al.*, 1969). In the Netherlands several cemeteries were surrounded by barberry hedges to keep the cattle out, and the naughty boys. Around these cemeteries black rust of rye occurred regularly (Oort, 1941).

Box 2.2. Sir Joseph Banks, the rust and the blight.

Banks (1805) wrote a small treatise on stem rust of wheat, published as an annex to a booklet on the cultivation of meadow and lawn grasses (Curtis, 1805). Both were illustrated by beautiful coloured plates. Following the disastrous wheat harvest of 1804 he studied the 'blight in corn [= wheat] occasioned by the growth of a minute parasitic fungus or mushroom on the leaves, stems, and glumes of the living plant'. He saw each teliospore on its pedicel as an individual plantlet.

Banks called the uredinial stage 'rust' and the telial stage 'blight'. He used the microscope to study the 'blight', but there is no evidence of experimentation. Banks touched upon several points of interest to the epidemiologist.

Quoting the farmers' belief, 'scarcely credited by botanists', in a relationship between the barberry and the 'blight' Banks speculated on the biological process underlying that relationship, without coming to a firm conclusion[72].

Banks mentioned a relationship between the 'chocolate-brown' rust and the black blight, that could cohabitate in one pustule, but he preferred to stay on the safe side and, following Fontana (1767), he considered them to be different species. He stated that the 'rust' was 'believed to begin early in the spring, and first to appear on the leaves of wheat in the form of rust, or orange-coloured powder'. I surmise that this early rust was yellow rust, rather than brown or black rust that tend to show up at a later stage of development.

'The chocolate-coloured Blight is little observed till the corn is approaching very nearly to ripeness; it appears then in the fields in spots, which increase very rapidly in size, and are in calm weather somewhat circular, as if the disease took its origin from a central position'. The late appearance, the colour, and the conspicuous focus formation are characteristic for black rust.

Banks believed the rust and the blight to be dispersed by wind, referring to the spores ejected by puff ball fungi. He was impressed by the capacity for multiplication of the rust and the rapidity of the malady's spread. Banks had a feeling for what we now indicate as temperature dependent latency period; 'at this season [spring], the fungus will, in all probability, require as many weeks for its progress from infancy to puberty as it does days during the heats of autumn'.

In 1848 black rust was known under various names such as *Uredo linearis* for the uredinial stage and *P. graminis* for the telial stage[73]. The aecial stage on barberry was seen as a separate rust species (*Aecidium berberidis*), though Banks and others had their suspicions. The host alternation from barberry to rye or wheat was known by several non-academic practicians since the early 19th century but *in academia* that relationship was ignored until it was conclusively demonstrated by De Bary (1866).

Westendorp, a medical practitioner, was active in Belgium as an amateur mycologist of high standing. In his commercial herbarium (Westendorp & Wallays, 1846) we find black rust in the uredinial stage[74] as *Uredo linearis* Pers. (Part 7 No. 331; 'on the stems, the sheaths and the leaves of cereals, mainly wheat') and the telial stage[75] as *Puccinia graminis* Pers. (Part 2 No. 91).

Brown leaf rust of rye was ill known as a separate species around 1850. Berkeley (1854/7) did not mention brown rust in his 'Vegetable Pathology', nor did De Bary[76] in 1853. Westendorp (1860) described[77] *Puccinia recondita* Rob. ex. Desm. on the wilting leaves of rye as a new record for Belgium. Scattered uredinial pustules of ~2mm occur primarily on the adaxial side of the rye leaves, also on the abaxial side, but quite seldom on leaf sheaths, stems, and heads. Their colour changes in the course of time from yellow-brown to chestnut. The host alternation with *Echium* and *Anchusa* spp was unknown until De Bary (1867). The brown rusts of rye and wheat look similar but they are different species[78]. The literature abounds with confusion in naming the brown rusts and several authors attached the name *rubigo-vera* (see Box 2.3) to brown rust[79].

Historical evidence for a significant role of the alternate hosts is not available. We assume that brown rust of rye in the Netherlands behaved epidemiologically as brown rust of wheat (Chester, 1946; Zadoks & Leemans, 1984) and yellow rust (see below). Since the two brown rusts have higher temperature optima than yellow rust their overwintering will be less frequent and their epidemic upsurges in spring will be later[80]. Wind dispersal of *P. recondita* urediniospores over large distances may have occurred in the 19th century with its wide-spread rye cultivation but there is no evidence whatsoever.

Yellow stripe rust (Figures 2.1 and 2.2) on rye seems to have been fairly general in the old days. De Candolle (1815) described the uredo rust[81] of cereals, *Uredo rubigo-vera* (see Box 2.3). This rust 'is born on the upper surface of the leaves, more rarely on the lower surface, on the sheath of the leaves or on the stem of the gramineous plants, and primarily on wheat; there she forms extraordinarily small oval pustules, but normally very numerous'. They produce a 'yellow dust: in the end, that dust becomes reddish, but never black'. 'The ovoid spores are nearly round'. The description excludes black rust, and typically fits yellow rust with the term 'yellow dust' and the superlatives 'extraordinarily small' and 'very numerous'. Nevertheless, brown rust[82] is not completely excluded because in the end the dust becomes 'reddish'. We meet with this ambiguity[83] in several descriptions as long as brown rust is not seen as a species *per se*.

De Candolle added that the rust, when abundant, exhausts the plants and causes considerable damage. Berkeley (1856: §601, 95) stated that *U. rubigo-vera* could have disastrous effects on wheat when 'the spikelets and the germen are affected'. Meyen (1841: 134) wrote that *U. Rubigo-vera* Dec. in many years was exceptionally frequent and ravaged whole counties. He proceeded (*ibid.*: 140) with a rather detailed description of glume rust (*Rubigo glumarum*), which he apparently considered to be another species than *rubigo-vera*. The comments by Berkeley and Meyen point to yellow rust, since brown rust on glumes is extremely rare.

Interestingly, the *Uredo rubigo-vera* Dec. (1846 Part 5 No 231) of Westendorp & Wallays on cereals was neither black nor brown rust, but yellow rust on rye (Figures 2.3 to 2.5a). As part

---

Box 2.3. Rubigo-vera – a nomen confusum.

For an epidemiological interpretation of widely scattered data a first requirement is the identity of the rust under consideration. To that purpose rust species have received names. Unfortunately, the early rust taxonomy is utterly confusing, see e.g. the historical paragraphs in Eriksson & Henning (1896: 143/5, 211/2). This box gives a few highlights[84].

Persoon (1801) was convinced that *Uredo* and *Puccinia* were different genera. He rejected the idea of a transition from the one into the other. 'Observation: I fear that [*V. linearis*] is not but the younger plantlet of *P. graminis*)'[85]. Most scientists followed his view, among whom Banks (1805), Fries[86] (1829), and Léveillé (1848). De Bary (1866) produced convincing experimental evidence to the contrary, describing macrocyclic and heteroecious rusts.

Erroneous splitting was accompanied by wrong lumping. Decandolle (1807) introduced the name *rubigo-vera* for rust on cereals in the uredinial phase. In Léveillé's detailed description (1848: 777) *rubigo vera* could infect leaves, sheaths, stems, glumes and even grains of cereals. He recognised two spore forms, a large oval one and a smaller roundish one, stating that the large oval ones did not really belong to the Rust (F: *la Rouille*). Thus he eliminated black rust from the complex. He knew that *rubigo vera* was a rust that clearly changed colour in the course of time[87]. He also knew that an attack began with oval dots, sometimes scattered (brown rust?), sometimes crowded (yellow rust?). He did not conclude that he dealt with a complex of two rusts.

The final definition of 'brown rust' (*Puccinia dispersa* nov. spec.) was made by Eriksson & Henning as late as 1896. Describing two *formae speciales*, *tritici* and *secalis*, these authors introduced a new mistake. Presently the brown rust on wheat is *triticina*, a name dating from 1899, and that on rye is *recondita*, a name dating from 1857 (Desmazières, 1857 p798[88]; Savile, 1984).

The original description of *rubigo vera* by Decandolle (1807), as quoted by Seringe (1818), perfectly fits the yellow rust. I could not examine his herbarium with the specimen on rye but I did examine the herbarium of Westendorp & Wallays (1850, Volume 5. No.231. *Uredo rubigo-vera* Dec.). I convinced myself that their specimen of *rubigo vera* was yellow rust on rye. This is the 'yellow rust line' of *rubigo vera* interpretation.

Meanwhile a 'brown rust line' of interpretation came into being. Eriksson & Henning (1896: 211/2) and others mentioned *rubigo vera* in the context of *triticina*, brown rust of wheat (*vide* Wiese, 1977: 39).

The name *glumarum* was given to a rust, apparently yellow rust, by Schmidt (1817, *ex* Eriksson & Henning, 1896), but I had no access to the source. Westendorp (1854b), in Desmazières' trail, used that name, but for what? Léveillé (1848) treated *glumarum* as a separate species, 'which has much analogy with the Rust, on glumes of wheat and rye, causing incidental abortion'. The phrasing does not suggest that Léveillé had seen Desmazières' *glumarum*[89]. The 'abortion' evidently referred to rye only. Eriksson & Henning (1896) finally coupled *glumarum* to yellow rust (G: *Gelbrost*), *Puccinia glumarum* nov. spec.

The rust epidemic of 1846 occurred in a period of taxonomic vagueness, so that we cannot trust names but must look for symptoms, often at field level, such as early appearance.

---

5 was published in 1846 we may believe that the herbarium sample was collected during the epidemic of yellow rust on rye in that year. The specimen[90] consists of two pieces, each a fragment of a stem with a complete leaf and part of a leaf sheath. 'On the leaves of cereals' says the text, without indication of host species, location, or date of collection. Glumes were shown nor mentioned.

Figure 2.1. Puccinia glumarum *(Schm.) Eriks. & Henn. – Yellow stripe rust, now* P. striiformis *Westend. From plate VI in Eriksson & Henning (1896). No rye depicted.*

66-72. Wheat.

66. Summer leaves, rust development in the course of 13 days.

67. Leaves in secondary stage of disease.

68. Leaves with dying rust, flecked.

71. Leaf sheath (x1), uredinia and telia.

72. Leaf sheath (x2), a - younger, b - older stage.

73. Barley, summer leaf, primary disease stage.

Figure 2.2. Puccinia glumarum *(Schm.) Eriks. & Henn. – Yellow stripe rust, now* P. striiformis *Westend. From plate VII in Eriksson & Henning (1896).*
74-80. Wheat.
74. Spike (x1).
75. Spikelet (x2).
76. Flower, a - glume, b - lemma, c – palea, d – outer and e - inner side of kernel.
77. Spikelet, awned (x2).
78. Spikelet (x2), uredinia and telia.
79. Glume (x2)
80. and lemma (x2), seen from inside, with uredinia and telia.

Desmazières[91] called the rust on the inflorescences *U. glumarum*[92]. Wiegmann's (1839: 115) description of *Uredo Glumarum* (G: *Spelzenrost, Kappenbrand, Balgbrand*) points to yellow rust. Westendorp (1854b) briefly mentioned *Uredo glumarum* Rob. ex Desm. The critical Berkeley (1856: §602, 788) wrote 'It is right to state that Desmazières considers the species which

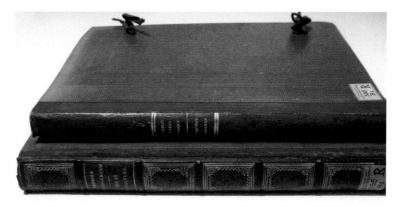

Figure 2.3. Two volumes of the Herbier cryptogamique ou collection des plantes cryptogames qui croissent en Belgique *(Cryptogamic herbarium or collection of cryptogamic plants that grow in Belgium), by G.D. Westendorp and A.C.F. Wallays, Gent, van Doosselaere. Present in the Library of the Wageningen University & Research Centre. Book sizes about quarto.*

grows on the inflorescence as different from that which grows on the leaves, and therefore has published it as *U. glumarum* Roberge; but after an examination of his own specimens this opinion is not borne out, for there is no one essential character to separate them, nor are the spores as he describes them smooth, but minutely granulated as in undoubted *U. rubigo-vera*'[93]. So, Berkeley's *rubigo-vera*, possibly ambiguous, at least included our yellow rust.

In fungal taxonomy the step from anamorph (asexual stage, here urediniospores) to teleomorph (sexual stage, here teleutospores) is important. Westendorp (1854a) described *Uredo striaeformis* nov. sp. on grasses[94] such as *Holcus lanatus*, on which I never found yellow rust. His description included the telial stage, though the word 'description' is too honorific[95]. The combined volumes 21 and 22 of Westendorp & Wallays (1855) contain N° 1077, *Puccinia striaeformis* West, 'on the straw of cereals around Courtrai' (Figure 2.5b). Macroscopic inspection showed the telial stage of yellow stripe rust[96]. The species name, adjusted to *striiformis*, received priority because it referred to the perfect stage (teleomorph) of the fungus (Hylander *et al.*, 1953)[97].

Finally, Eriksson & Henning described *P. glumarum* as a separate species. They (1896: 203) considered *P. glumarum* f.sp. *secalis* on rye to be a separate *forma specialis*, morphologically identical with the *ff. sp. tritici* and *hordei* on wheat and barley, respectively. They used, maybe introduced (1896: 141), the name 'yellow rust' (G: *Gelbrost*). I have no reason to believe that the ecological behaviour of the f.sp *secalis* differed from that of the other two *formae speciales*, though I don't know for sure. Yellow rust has no known alternate host.

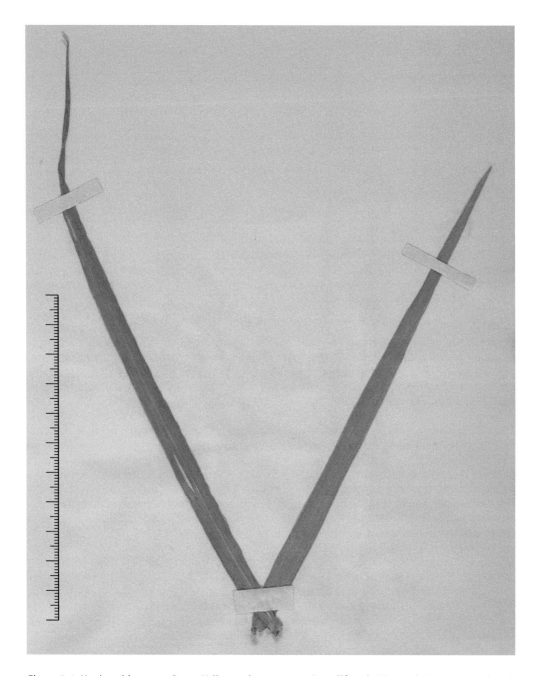

Figure 2.4. Uredo rubigo-vera *Dec. - Yellow stripe rust, now* P. striiformis *Westend. From Westendorp &
Wallays (1846),* Herbier cryptogamique ou collection de plantes cryptogames et agames qui croissent en
Belgique *(Cryptogamic herbal or collection of cryptogamic and agamic plants that grow in Belgium), Brugge,
de Pachtere, Vol. 5, No. 231.*
 *'On the leaves of cereals. Received from Mr. L. Matthys'. The specimens are most probably rye, left the abaxial
and right the adaxial leaf side with the uredinial stage only. Possibly collected during the 1846 epidemic.*

a.
b.

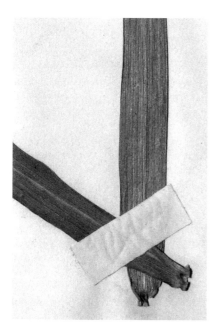
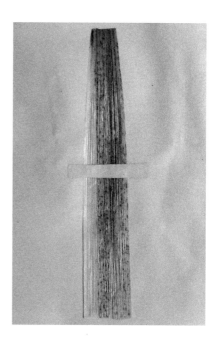

Figure 2.5a. Uredo rubigo-vera *Dec. – Detail from Figure 2.4, adaxial leaf side; b.* Puccinia striaeformis *West. – Yellow stripe rust, now* P. striiformis *Westend. From Westendorp & Wallays (1855), Vol.21/22, No. 1077. On the culms of cereals around Courtrai. Probably on wheat, telial stage, no uredinia seen. The small telia are arranged in stripes, shiny dark brown to black.*

Vernacular names of plant diseases in the 19[98] century are a mess[98]. In Zeeland (SW Netherlands), 1957, an elderly farmer mentioned the name 'honeydew' (see Box 2.4) to me for yellow rust, a rather non-specific term from the 19[99] century[99]. In the Dutch literature of the first half of the 19th century we find on rye the Dutch equivalents of 'red-dog', 'red rust' or 'the red' (Roessingh & Schaars, 1996: 446). In my experience 'red', as used by farmers, is a non-specific term rather meaning 'not-green'[100]. However, the uredinia of brown rust are reddish brown and those of black rust are somewhat darker fox-brown or chocolate-brown. The name 'red-dog' may have been a reminder of classical writings (Chapter 1). Van Hall (1828: 314) quoted an informant writing on 'the red' (*Uredo Fabae* Pers.) of broad beans[101], thus providing a direct link between the names 'red' and 'rust', a link he repeated in 1854 (284).

Box 2.4. Honeydew (= D: honingdauw or honigdaauw, G: Honigthau).

In 1957, when I studied the epidemiology of yellow rust on wheat in the field (Zadoks, 1961), an elderly farmer in Zeeland told me 'formerly we used to call this honeydew'. As a vernacular name for yellow rust the word 'honeydew' is legitimate though confusing. The old man's comment led me to the honeydew controversy. Around 1800, parasites such as rust were considered to be the cause of disease by some, and the consequence of disease by others.

Cause. Fontana (1767), Banks (1805) and De Candolle (1815) were convinced that parasites were 'plant' species living separately from but housed on or even in their host plants. They were sitting – as it were - at the same table, the original meaning of the Greek work παρασιτος (= parasite, sitting by). Clearly, Fontana (1767: 34) realised that these independent plants were living at the cost of their host plant, the present meaning of the word parasite (*ibid.*: 34). So did Banks and De Candolle.

Consequence. The other party had reminiscences of the humoral theory, of classical Greek origin and in accordance with 18th century medicine. Possibly they were troubled by Fabricius' (1774: 36) fuzzy discussion of 'Honey-dew, *Erysiphe*'. Unger's (1833) exanthematic theory was the apogee of that line of thought in plant pathology, and also nearly the end. Plants in a poor condition exuded humors that 'cristallised' into a shiny substance which today we call honeydew, or materialised into one or another fungus. Unger described and named these fungi in the usual way, provided nice drawings of fungi rooting in the host plant, but nevertheless considered them explicitly as the consequence of disease. Wiegmann (1839) and others followed Unger's line of thought[102].

Van Hall (1828: 307), usually well informed, had read that insects could produce a substance, called 'honeydew', that precipitated on the leaves of plants but he was not fully convinced. He knew that honeydew was on and rust in the leaf. He thought it improbable that honeydew and rust were successive stages of the same disease, since he had observed honeydew without subsequent rust and rust without preceding honeydew. However, in 1854 Van Hall (1854: 284) equated honeydew and mildew (*Erysiphe*), being unaware of the fact that the old root 'mil' means honey and not meal. Van Hall returned to the subject in 1859, evidently confused by Unger's (1833) exanthema theory, and then suggested that honeydew was an early phase of rust. Which rust?

Enklaar (1850: 177) wrote 'The honeydew is the sweated, sugary juice of the crops'. He connected honeydew and aphids, though in an inverted sense, first the honeydew and then the aphids.

The learned Dutchmen apparently overlooked a comment by Erasmus Darwin (1800: 326) 'In a paper written by Abbé Boissier de Sauvages, he describes two kinds of honey-dew; one of which he concludes to be an exudation of the tree, and the other he asserts to be the excrement of one kind of aphis, which the animal projects to the distance of some inches from its body on the leaves and ground beneath it; and which he believes the animal acquires by piercing the sap-vessels of the leaf. The circumstances are distinctly described, and by so great a philosopher[103] as Sauvages of Montpellier, that it is difficult to doubt the authenticity of the fact.' Darwin was not convinced: 'But that a material so nutritive should be produced as the excrement of an insect is so totally contrary to the strongest analogy, that it may nevertheless be suspected to be a morbid exudation from the tree; ...'

2.2.3. Epidemiologic phases

A severe epidemic of yellow rust on wheat passes through several phases (Zadoks, 1961). Phase 1 is a wet summer with many volunteer plants permitting an ample *oversummering* of the rust. Phase 2 is a fair and dry autumn which many early-sown winter crops allowing the *switchover* of the rust from the preceding to the following crop[104]. Phase 3 is a mild winter enabling the rust to settle in the new crops and to multiply repeatedly with comparatively short generation periods, the *overwintering*. Phase 4 is a mild spring during which the rust multiplies at a faster rate with increasing temperatures so that it 'overtakes' the rapidly growing crop, the *outburst* of the rust. Supposedly the four stages apply also to yellow rust on rye in the 19th and 20th centuries.

Good crop rotation reduces the risks of over-summering and switchover (Zadoks, 1961). Unfortunately, the rye farmers on the sandy soils – especially those in the East of the Netherlands – about the mid 19th century did not bother much about crop rotation, frequently growing rye after rye (Van Zanden, 1985: 186).

Urediniospores of yellow rust can be wind-dispersed over hundreds of kilometres but these wind-borne spores rarely initiate a typical outburst in the season of dispersal. A heavy epidemic has a local origin and requires a build-up of a year (the four phases). As soon as the temperature reaches a certain limit (~20 °C) the yellow rust epidemic comes to a stop whereas the brown and black rust epidemics continue to grow.

Several authors mentioned that yellow rust could be more severe in well-fertilised crops, as is my experience in the late fifties with nitrogen dosage experiments in wheat. The most amusing remark was on the corn (read rye) grown at Altona (then Denmark) near Hamburg (Germany). Rust became frequent when farmers began manuring their crops with 'herring' (Enklaar, 1847). The waste of the herring industry provided cheap organic matter, rich in nitrogen.

2.3. The rust on rye in 1846

Retrospective identification of a disease agent is tricky. Fortunately, we find support in the literature. Our primary source is the 'General Report on the State of Agriculture in the Kingdom of the Netherlands'[105]. This report collated the data provided by provincial informants. The point is relevant since every province had its own informant(s). When a specific comment, e.g. a symptom, is observed in two provinces we have – in principle – two independent and mutually reinforcing pieces of evidence. The reservation 'in principle' refers to the possibility that two informants discussed the symptoms before reporting.

In Appendix 2.1 the relevant sections are reproduced in bold and analysed. The Dutch has not been translated into English as translation may introduce interpretation. The analysis is organised by province. Quotes are [numbered] for further discussion. Arguments for and against the conclusion 'yellow rust' are [lettered], written in English and arranged in order of persuasiveness.

The conclusion is 'yellow rust', with the possibility of incidental 'brown rust' and with exclusion of 'black rust'. The main argument is the yellow-coloured infection of the glumes. A characteristic observation was made by Anonymous (1846b: 47). The cloths of a person walking through the crop turned yellow of the heavy fungal dust (conform author's experience), and dogs entering the crop came out yellow. None of the observations on the infestation, the course of the epidemic, and the environmental conditions contradict the conclusion.

The evidence produced in this chapter in support of the conclusion 'yellow rust' is based on epidemiological phenomena, macroscopic symptoms, and observations at the level of the hand lens. Observations under the microscope were rare and mainly referred to the roundish shape of the urediniospores, a characteristic shared by the yellow and brown rusts, thus excluding black rust. One observer mentioned that the spore walls were hyaline (Göppert, 1846), thus excluding brown rust[106].

2.3.1. The weather in Europe

The very mild winter 1845/6 and the favourable months of March and April were followed by an exceptionally hot and dry summer in most of Europe, from the Atlantic to the Black Sea, from the Netherlands to the Alps, even S to Sicily (Bourke & Lamb, 1993: Chapter 3). A variety of crops, primarily spring-sown crops, was damaged by heat and/or drought. Rye, normally fall-sown, was the preferred crop for the poorer soils that were drought-sensitive. Drought damage in rye occurred from France to Poland.

In Poland and Silesia strong, cold winds from N and E directions damaged early winter rye when about to bloom, hampering fertilisation and seed set. Large stretches of rye had completely or partly deaf ears (Weber, 1847). Damaging night frosts must have occurred but the available information is scanty and imprecise[107].

Yellow rust can progress without rain as long as there is ample nightly dew (Zadoks, 1961). Many records stated that the atmosphere was often humid, even oppressive. Under such conditions, given clear night skies, dew is guaranteed, certainly in relatively low lying areas.

2.3.2. The Netherlands[108]

Naming the rust. Ponse (1810) wrote about honeydew as occurring mainly on wheat, especially on late ripening wheat. In 1827 he responded to a price contest on honeydew. Evidently, honeydew here stands for rust. Ponse followed Banks (1805). Some details suggest that he dealt with yellow rust which gradually 'turned into' brown rust toward the end of the season. His honeydew began early in spring, with excrescences of an orange colour, 'and appears in the form of rust, of orange-coloured dust', as yellow rust does. 'Chocolat-coloured honeydew is rarely detected before the grain approaches its full ripeness, appears in tiny flat surfaces, somewhat roundish when the weather is calm'. The term 'chocolat-coloured' comes straight from Banks who used it for the colour of black rust uredinia. However, Ponse probably saw uredinia[109] of brown rust, which may indeed appear as oval 'facets' (Ponse's term, D: *ovale vlakjes*), a description not appropriate for black rust uredinia. Ponse used the word 'rust' (Banks's term for the uredial stage of black rust), saw both yellow and brown rust, but did not realise that the symptoms seen belonged to different rust species[110].

Enklaar (1850: 177) used the terms 'rust' en 'honeydew' in their modern sense but he did not discuss rust species. He saw '... yellow, later rust-coloured flecks and stripes on the leaves and stems, sometimes also in the ears of the cereals ...', mainly on wheat and rye. The fluent colour transition reminds us of Ponse. The words 'yellow', 'stripes' and 'ears' suggest yellow rust. He continued 'thus the rye harvest of 1846 was destroyed for a large part by rust'. Anonymous (1846b,c) described the epidemic in detail. He concluded correctly '*Uredo vera* Decandolle', first forgetting the '*rubigo-*' (1846b: 48) but adding it later (1846c: 99).

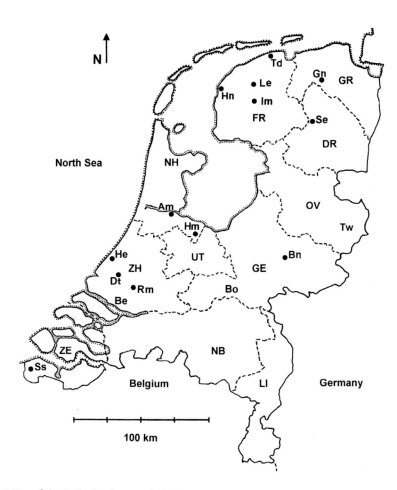

Figure 2.6. Map of the Netherlands around 1846.

XY = Province		Am	Amsterdam
DR	Drenthe	Bn	Brummen
FL	Flevoland	Dt	Delft
FR	Friesland	Gn	Groningen (town)
GE	Gelderland	Hn	Harlingen
GR	Groningen (province)	He	The Hague
LI	Limburg	Hm	Hilversum
NB	Noord-Brabant (shortly Brabant)	Im	Irnsum
NH	Noord-Holland	Le	Leeuwarden
OV	Overijssel	Rm	Rotterdam
UT	Utrecht	Se	Smilde
ZE	Zeeland	Ss	Sluis
ZH	Zuid-Holland	Td	Ternaard
		Be	Beijerland
		Bo	Bommelerwaard
		Tw	Twente (E part of Overijssel)

Van Hall (1854: 284) described rust on cereals. His *Uredo Robigo* (early infection in lower leaves, later stems and ears attacked, orange-coloured spore mass) was certainly not brown or black rust; a poor figure (Figure 2.7) showed the typical stripes of yellow rust. 'Some years earlier the rye harvest failed nearly completely', he wrote without mentioning 1846 explicitly. As late as 1859 Van Hall adhered to the exanthematic theory[111] of the origin of rust and mildew.

Figure 2.7. Uredo rubigo. Figure 286 in Van Hall (1854). 'Some years ago the rye harvest herewith failed nearly completely'. ('Voor eenige jaren is hierdoor de oogst der rogge bijna geheel mislukt').

Winter 1844/5. The winter 1844/5 had been exceptionally severe and long with much snow (Easton, 1928: 151; Figuur 3.1). Many frozen winter crops were ploughed under to be reseeded in the spring of 1845. Most rust inoculum will have disappeared but in places, some may have survived under the snow cover (Zadoks, 1961). The 1845 harvest season must have been late due to the long winter and, if so, the lateness favoured the *oversummering* and *switchover* of the rust.

Summer 1845. The summer of 1845 had been wet as confirmed by the reports of some provincial governors (Rüter, 1950):
- 'In many regions the rye is germinating [= sprouting in the ear] due to wetness' (Overijssel, 21 August: 400).
- 'The rye suffered from prolonged rains, but harvest rather sufficient' (Zeeland, 4 September: 403).
- 'Corn [= rye] lodged due to frequent gusts of wind and rain' (Noord-Holland, 29 August: 434).
- 'Rye suffered by germination leading to quality loss' (Gelderland, district Zutphen, 5 September: 441).
- 'Rye harvest better than average, harvest much retarded by continuous rains, quality much reduced' (Drenthe, 1 September: 454).
- 'Rye and wheat suffered much from wet weather and strong winds' (Friesland, 8 September: 454).

This contemporaneous information is the best proxy available for *oversummering* and *switchover* of yellow rust. The 'General Report' over 1845 did not mention rust on rye. Apparently the over-summering and switchover had not been noticed.

Winter 1845/6. The winter of 1845/6 was very mild. The relatively high winter temperatures must have facilitated the *overwintering* of yellow rust. No reference was found to direct observation of the rust.

Spring and summer 1846. The *outburst* of the rust on rye was first seen in Gelderland (mid April; Anonymous, 1846b: 46), in Brabant (end of April; General Report over 1846: 28), in Groningen as of May 1st (*ibid.*: 40), and in Noord-Holland in the course of May (*ibid.*: 33). These 'firsts' are relatively late. Noteworthy is the observation of rust on winter barley in Groningen in March and April (*ibid.*: 53), 'apparently the same species which was later so common on rye'. March and April cover the period of the early outburst of two rusts on winter barley, the yellow rust (*P. striiformis* f.sp. *hordei*) and the brown rust of barley (*P. hordei*). The words 'same species' allow excluding the latter rust.

The First National Congress of Agriculture[112], convening in June, 1846, discussed – among more important matter – the situation of the corn [= rye]. Delegates tabled rusted rye culms for examination and discussion. They were not farmers, no, they 'had' farmers, but their interest was genuine. Some illuminated men called the rust on leaves and stems *Vredo rubigo*, omitting De Candolle and his *vera*. But the rust on the heads, was it the same or a different one? In the E part of the country expected losses were estimated at two thirds of a normal harvest. In Friesland the rust was still spreading, from sandy to clay areas. At the second Congress, June, 1847, the worthy gentlemen looked forward to a bumper crop of rye; the pollen clouds had been exceptionally profuse in 1847. Only then the participants remembered that the pollen clouds of the rye in 1846 had been miserable, with the implication of a poor fruit set in that calamitous year (Appendix 2.1, comment [30]).

The Governor of the Province Noord-Holland reported 'Due to the rust, that made itself known already on the leaves of the rye, and later spread to the lower part of the stem [i.e. leaf sheaths?] and on to the heads, this crop did not produce more than half of the usual yield'(Van Ewyck, 1847). The same is approximately true for the national rye yield (Figure 2.8). Rye prices soared and extra imports were needed (Figures 2.9 and 2.10).

The losses in 1846 were high, roughly about half of the expected yield (see Box 2.5). Rust on rye and wheat occurred regularly in the Netherlands during the 19th century. Kops (1808: 81/4, 149) provides an example. After the mild winter 1806/7 rust or honeydew occurred in various provinces. In rye this was 'red rust' (Groningen), 'honeydew' (Gelderland) and 'the red' (Drenthe). The lateness of observation, June (Groningen) and 'after the flowering of the winter rye' (Drenthe) points to brown rust though it does not exclude yellow rust.

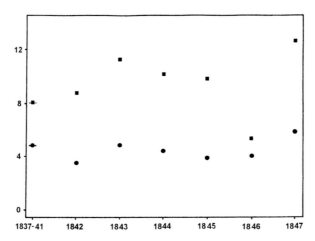

Figure 2.8. National wheat and rye production of the Netherlands, 1837-1847.
Horizontal – years. Vertical – yield in units of 100,000 last (one rye last = 2100 kg). ■ = rye, ● = wheat. The two crops react roughly similar to the vagaries of the weather. The low rye yield in 1846 must be due to a differential factor, here yellow stripe rust. The 1847 yields were quite good. Original data from Ackersdijck et al. (1850).

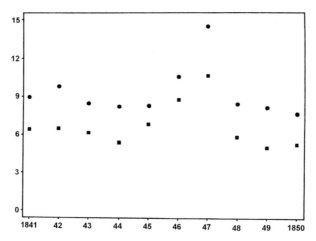

Figure 2.9. Prices of rye and wheat in the Netherlands, 1841-1850.
Horizontal – years. Vertical – price in Dutch guilders per hl. ■ = rye, ● = wheat. After Staring (1860: 586).

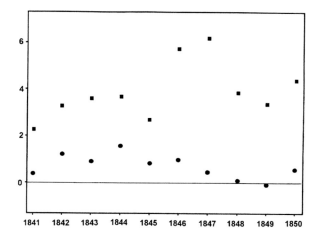

Figure 2.10. Net imports of wheat and rye in the Netherlands, 1841-1850.
Horizontal – years. Vertical – net imports in units of 10,000 last (one rye last = 2100 kg). ■ = rye, ● = wheat.
Imports of rye peaked in the years 1846 and 1847. Original data in Ackersdijck et al. (1851: 243).

Box 2.5. Yield losses of rye in the Netherlands, 1846, by province (See also Appendix 2.1).

Drenthe. Bieleman (1987, graphs 6.16, 6.17) provided a long-term overview of rye yields in that province. Yields were quite variable, for a variety of reasons. The rye yields of 1846 were ~50 per cent of the mean for the yields in 1845 and 1847. No explanation was given. I presume that most of the loss has to be ascribed to yellow rust.

The 'General Report' provided some scattered information according to province:
Friesland. Village Ternaard, some fields with total loss.
Groningen. 50 to 67 per cent loss relative to expected yields.
Drenthe. Village Smilde, ~50 per cent loss (see also above).
Gelderland. ~67 per cent loss. Wttewaal (1848) ascribed the loss to *Uredo Rubigo vera* De Cand. on the leaves and ears and to *Puccinia graminis* Pers. on the ears. He may have seen the telial stage of *P. striiformis* on the ears, at the time not yet described.
Noord-Holland. Loss locally over 50, at the village Hilversum up to 60 per cent.
Zuid-Limburg. Loss often 100 per cent, produce of many fields not threshed but fed to sheep.

At the First National Congress of Agriculture, 1846, loss estimates were reported (Anonymous, 1846/7):
Overijssel. Region Twente (E part of the province), ~67 per cent loss.
Gelderland. Village Brummen, ~67 per cent loss.

The usual comment was that the quality of the bagged rye was excellent.

The aftermath. The 'General Report' over 1847 mentioned rust on rye in North Holland ('some fields'), Gelderland ('not on the kernel'!) en Groningen ('no damage to the kernel'!). The 1847 harvest was excellent, without losses. Wttewaall (1848) stated that *Uredo rubigo vera* Dec, which had caused so much destruction in 1846, had not yet departed in 1847, though in the General Report he mentioned the presence of some foliar rust in the second half of June. In June, 1848, the rust was noticed in the provinces of Gelderland and Overijssel, but it was too late to cause damage. Apparently, people were on the alert[113]. It seems as if the epidemic of yellow rust on rye vanished as fast as it appeared. However, severe yield losses in Drenthe, 1859, have been associated with 'red rust' (*Puccinia rubigo vera*?; Bieleman, 1987: 655).

2.3.3. Other countries in Europe[114]

Eriksson & Henning (1896) did not see morphological differences between f.sp. *tritici* and f.sp. *secalis* and they did not provide experimental evidence of a physiological difference between the two *formae speciales*. They must have seen that sometimes the wheat and sometimes the rye was infected, and that the rust apparently did not move from wheat to rye or *vice versa*[115]. They had read Bjerkander's[116] (1794) description of severe attacks by yellow rust on rye in Sweden. He wrote that the ears became yellow leading to formidable damage; nothing about wheat. German data from 1846 recognised differences in rust appearance and severity between wheat and rye. This implies that the difference between the two *formae speciales* existed already.

The climatic conditions in most of W Europe were comparable to those in the Netherlands, an unusually severe and long winter 1844/5 with heavy snow cover, and a very mild winter 1845/6 (Easton, 1928: 151/2).

In **Belgium**, where rye was the dominant[117] arable crop, great damage by rust occurred in 1846. Vanhaute (1992) wrote that in the Campina ('Kempen', the sandy part in the north) half of the rye harvest had been lost without indicating a cause. Was it by frost, drought, or rust? In Flanders, 1846, rye yielded 7 in stead of 18 hl/ha (Lamberty, 1949: 136). Lamberty did not mention a cause, but Pirenne (1932: 128) wrote 'rust'. Treviranus[118] (1846) had travelled through Belgium from 20 till 26 April and saw a very severe rust attack on rye, apparently yellow rust. So far this is the only written indication that in Belgium yellow rust was the culprit.

Appendix 2.2 gives an overview of harvest data from Belgium. The high grain losses in combination with a good quality (expressed as hl weight) remind us of the information from the Netherlands and Germany about the combination of severe loss and good quality that is typical for yellow rust on rye. Pirenne's rust must have been yellow rust. The seriousness of the situation is demonstrated by the net imports of rye into Belgium, which tripled during the emergency years 1846 and 1847, whereas import prices nearly doubled (Figure 2.12). Telia were not mentioned in 1846 though they must have been present.

Denmark. The *Nieuwe Rotterdamsche Courant* (July 25th, 1846) reported from Copenhagen that people from nearly all regions complained about the rust in rye though they still hoped that the rust would be of little consequence. Rye yielded about half the expected amounts due to drought and, probably, rust (Löbe, 1847: 134). In Holstein, then Danish, rye yielded about one third of the expected amount (Weber, 1847: 14).

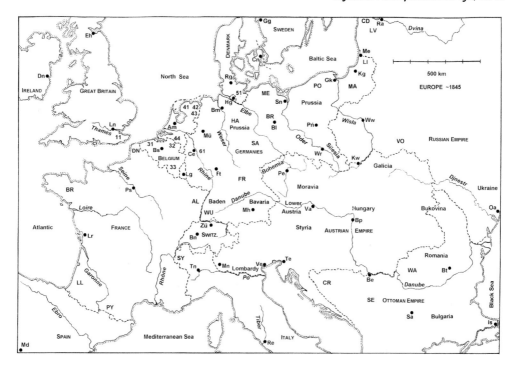

Figure 2.11. Map of Europe, situation of ~1845.
In the W we see the rather stable nations Great Britain, France and Spain. The E is covered by the Russian, Ottoman and Austrian Empires. Going from N to S in the central zone we see Sweden and Denmark, 'the Germanies', future Switzerland, and future Italy. 'The Germanies' comprised numerous states, among which Prussia (the largest and most powerful), Bavaria and Baden. Prussia reached roughly from the Netherlands in the W to the Russian Empire in the E. The Kingdom of Poland was de facto incorporated in the Russian Empire. The Austrian Empire comprised present Austria, Hungary, Czech Republic, Slovakia, and parts of present Poland, Ukraine, Romania, Serbia, Croatia, and N Italy. After Westermann (1956, maps 126 and 127, Europe 1815, the peace arrangement by the Congress of Vienna).

Letter code: Countries, Areas, Seas, *Rivers*

11 Kent	DN Département	Am Amsterdam	Is Istanbul (=	Ps Paris
31 Flanders	Nord	Be Belgrade	Constantinople)	Ra Riga
32 Campine	FR Frankenland	Bl Berlin	Kg Kaliningrad (=	Re Rome
33 Wallonia	HA Hannover	Bm Bremen	Königsberg)	Rg Rendsburg
41 Friesland	LI Lithuania	Bn Bern	Kw Kraków	Sa Sofia
42 Groningen	LL Les Landes	Bp Budapest	Lg Luxemburg	Sn Stettin (=
43 Drenthe	LV Latvia	Bs Brussels	Ln London	Szczecin)
44 Noord-Brabant	MA Masuria	Bt Bucharest	Lr La Rochelle	Te Trieste
51 Schleswig-	ME Mecklenburg	Ce Cologne	Md Madrid	Tn Turin
Holstein	PO Pommerania	Cn Copenhagen	Me Memel (=	Va Vienna
61 Westfalia	PY Pyrenees	Dn Dublin	Klaipéda)	Ve Venice
AL Alsace	SA Saxonia	Eh Edinburgh	Mh Munich	Wr Wrocław (=
BR Brandenburg	SE Serbia	Ft Frankfurt	Mn Milan	Breslau)
BR Brittany	SY Savoy	Gg Göteborg	Mü Münster	Ww Warsaw
CD Courland	VO Volhynia	Gk Gdańsk (=	Oa Odessa	Zü Zürich
CR Croatia	WA Wallachia	Danzig)	Pe Prague	
	WU Württemberg	Hg Hamburg	Pń Poznań (= Posen)	

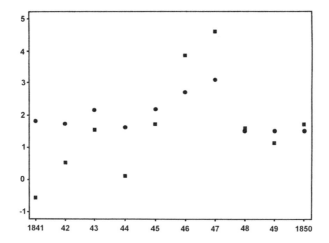

Figure 2.12. Imports of rye into Belgium, 1841-1850 (After Degreve, 1982).
■ = *Net imports of rye in units of 10,000 tonnes, transit excluded;* ● = *Mean import price of rye in units of 100 Belgian Francs per tonne. After Degreve (1982).*

France. Léveillé (1848), writing shortly after the 1846 epidemic, would have made a statement if yellow rust had been important in France. He didn't, though he knew about the devastation in Saxony. He knew that *U. glumarum* Rob. in Dsmz. occurred on the glumes of wheat and rye and sometimes caused partial abortion in rye[119]. Various French authors concur in calling 1846 a very bad year for cereals, primarily due to excessive drought (e.g. Jardin & Tudesq, 1973: 234; Chapter 3). The Alsacian rye harvest was not too good (see Box 2.6-2).

Germany. The loss data of the Kingdom Prussia were published early in 1847 (Table 2.1). No explanation was given. Wehler (1987: 643) stated that the 1846 yield of rye in Prussia was ~57 per cent of the mean yield in earlier years, without mentioning a cause for the dramatic loss implied in this figure. In Upper Silesia rust 'devastated large stretches' of rye[120]. The daily newspaper *Nieuwe Rotterdamsche Courant* (1846) quoted losses of ~50 per cent in the Pfaltz (July 11th), 75 per cent in the Lower Rhine Valley (July 13th, 24th), 50 to 75 per cent in Wetterau, the Lahn valley and the Bavarian Rheinpfaltz (July 24th), and recurrent rust problems near Hamburg (July 18th). The German daily newspaper *Allgemeine Zeitung* tried to present a more optimistic view, but between the lines a rather sad reality is seen (see Box 2.6).

Von Schlechtendal (1846) and De Bary (1853) used the name *glumarum* for the glume-rust[121]. Auerswald wrote to Léveillé about that 'darned pest' (*nefaria pestis*) in Saxony, 1846[122]. Göppert (Silesia, present S Poland) indicated that the rusts on rye and wheat were not identical[123]. In the extreme NE of Prussia, in Sambia, the rye was bad, with an unidentified ear disease – yellow rust?[124] Without hesitation Eriksson & Henning (1896: 432) ascribed this epidemic on rye to yellow rust.

A communication from Münster in Westfalia, dated June 10th, 1846, was submitted to the 46th session of the Prussian Academy of Sciences (on July 4th, 1846). It said '... the disease-like appearances, that before were observed only on the veritable leaves, now also proceeded to the leaf sheaths, the culm, even on the ears and the kernel, and that the same could be found

Table 2.1. Kingdom Prussia, 1846, yields per province in per cent of 'normal'.
Data (dated January 4th, 1847) were published in Anonymous (1847a: 400/1)[125] and Anonymous (1847c). For the present table three crops were selected. Yield loss in wheat probably represents the effect of the drought and the high temperatures. The drought may have been worst in the E as suggested by the wheat losses, high in the N, and modest in the W. Yield loss in rye probably represents the combined effect of drought and rust. Note the very low yield in the Rhine Province where the yellow rust epidemic rampaged. Losses in potatoes were due to drought, heat and late blight, with larger losses in the E than in the W, the opposite of the 1845 situation. In the E the heat damage to potatoes was worst (Chapter 3).

Province	Situation	Wheat	Rye	Potato
Prussia	E	74	62	33
Poznań	E	64	61	48
Pommerania	NE	79	64	30
Brandenburg	N	73	66	63
Silesia	SE	71	60	53
Saxonia	S	74	59	63
Westfalia	NW	81	53	62
Rhine Province	W	87	50	7
Kingdom Prussia		75	59	53

everywhere: no soil type, no cultivation method, no preceding crop seemed to make exception: the more the rye develops, the more the disease on it appears at a higher stage' (Anonymous[126], 1847b: 179). The symptom descriptions suggested either black or yellow rust. As the epidemic was general and early (a note from May 31st was mentioned) and not focal and related to the vicinity of barberry (Windt, 1806), I opt for yellow rust[127].

Braun (1846) described the glume-rust of rye, 1846, around Freiburg (Baden), where no rye field was free of rust and seeds were sometimes totally destroyed, with serious losses. Uredinia were arranged in stripes on the leaf disk, at the inner sides of the leaf sheaths, not on the stem, but regularly on the glumes; typically yellow rust. The damage ran up to over 50 per cent of the expected yield. In Rheinland-Pfalz, with Mainz as its capital and grain market, the rye harvest failed (Box 2.6-3). Treviranus (1846) reported heavy damage on rye in the Rhine valley, around Siegburg and in the dale of the river Ahre. He saw a relation between the location of the rye field and its rust severity, protected fields in the valleys having more but high and exposed fields less rust[128]. He did not see a relation between severity and soil type.

Braun and Treviranus were intrigued by the poor fertilisation of rye during the epidemic. Severity was so high that in one third to one half of the florets no anthers appeared and sometimes no stigmata either. The stigmata were dusted with yellow. Hardly one half of the florets showed good fruit setting[129]. The harvest was meager, often less than one half of the expected amount, but the grain quality was good[130]. Similar observations, a poor harvest but of good quality, were made in the Netherlands (Anonymous 1846c: 100; Appendix 2.1 [n]). 'Good flour' (Box 2.6-12) is to be read as an allusion to the frequent abortion in rye.

Box 2.6. Quotes from the Allgemeine Zeitung, 1846, the S German 'General Newspaper'

At the time, the German *Allgemeine Zeitung* (General Newspaper), Augsburg issue, was published by Cotta in Stuttgart. A predecessor of the present, well-read *Frankfurther Allgemeine Zeitung*, it had a political rather than commercial orientation. Being a newspaper for the ruling class, the affluent and the complacent, containing much international political news and pages on literature, arts, or travel, it took a rosy view of the world. Bad news was clad in veiled language, often hidden among the small lettering of market reports. Crops looked 'just fine'. Nonetheless, grain prices did not go down as they should at the approach of harvest. Merchants began to order grain from abroad. When harvest data became available, the catastrophe could no longer be hidden. [note: /#issue/day-month/page number.]

1. #202/21-07/1616. Breslau (Wroclaw, SW Poland) - grain prices remain high, rye harvest expected to be less than average.
2. #206/25-07/1644. Frankfurt am Main - poor rye harvest will be compensated by good wheat crop. Rye quality in Upper Alsace was poor.
3. #206/25-07/1648. Mainz - rye harvest poor, merchants order rye from Russia and the USA.
4. #212 /31-07/1694. Posen (Poznań, present W Poland) – winter rye harvest bad, spring rye harvest meagre, 'grain prices continued to be high, and the distress of the working classes still does not want to decrease'. (*Die Getreidepreise halten sich andauernd hoch, und die Noth der arbeitenden Classen will sich noch immer nicht mindern*).
5. #216/04-08/1728. Silesia – rye harvest finished (no comments meaning bad news).
6. #217/05-08/1732. Brussels – Belgian rye harvest ½ of normal, famine expected if potatoes fail.
7. #218/06-08/1743. Teplitz (Teplica in Czech Republic) – grain harvest fair, grain weight low.
8. #222/10-08/1775. Lower Banat and Walachia (Middle and Lower Danube plains) – yields low, locally severe damage by wood mice (probably the long-tailed field mouse, *Apodemus sylvaticus* L.).
9. #225/13-08/1800. Silesia (modern SW Poland) – Rye with much straw but little grain, yields down to 1/3 of normal.
10. #228/16-08/1824. Bavaria – rye light in weight. County Mark – rye diseased, yields ½ to 1/3 of normal, disease all along river Rhine from Cleve down to Bingen.
11. #230/18-08/ 1838. Posen (Poznań, present W Poland) – Harvest was less than medium. Rye harvests in Prussia, Poland and N parts of Austrian Empire generally poor.
12. #236/24-08/1888. At 1300-1900 m altitude rye nearly ripe, but rusted. 'Good corn will be rare, but certainly there will be many fields in many regions which yield quite poorly. Much, very much straw, but grain below average; good flour'. (*Volkommenes Korn wird selten seyn, wohl aber gibt es ganz gewiss in vielen Gegenden viele Aecker die ganz slecht ausgeben. Viel, sehr viel Stroh, im Korn untermittelmässig; mehl gut*). Yields averaged over the whole of the Germanies will be ~50 per cent (14/27) of normal.

The various communications quoted above were confirmed in the review papers by Löbe and Weber published in 1847 (Appendix 2.3). In Brunswick (N Germany) the rye was severely diseased, as in Pommerania (NE Germany). In Silesia (modern SW Poland) rust on rye was destructive over large areas, as in then Poland (modern E Poland). As the harvested grain was of good quality, we think the rust was yellow rust indeed. The observation of the hyaline spore walls by Göppert (1846), published in the Silesian Journal, supports this idea, that would also explain the disease of the rye ears in Sambia, on the Baltic Coast.

The reviews bring four points to the fore. (1) Drought did much damage to spring cereals and to fall-sown rye in the Austrian Empire, Germany, Denmark, and Poland. (2) The fates of winter wheat and winter rye, usually coupled, often diverged in 1846 due to drought and rust. (3) The rust was nearly ubiquitous in Continental Europe N of the Alpine ranges, with the exception of France. (4) The available information, with several mentions of low grain yield but excellent milling quality of the remaining grain, clearly points to yellow rust of rye. Point 4 does not exclude, however, the possibility that brown and black rust appeared occasionally.

In 1846 rust on wheat was common too. An unidentified rust, not brown rust but either yellow rust or black rust, ravaged wheat in Mecklenburg, N Germany (Pogge, 1893).

Switzerland. The head-gardener of Zürich College (Regel, 1854) knew his plants and his literature. He described[131] the 'yellowing' and the 'light fruit' of corn [= rye] due to *Uredo Rubigo vera*, that caused so much damage in his area (N Switzerland). Using the microscope he found some 9000 spores per sorus of some 2 mm long. This symptom is typical of yellow rust and excludes black rust.

2.3.4. Yellow rust pandemics

The general appearance of yellow rust on rye over a large area, Belgium, Germany, present Poland and the Netherlands, seems to contradict the local origin of a severe attack as indicated above. Similar large-scale, nearly pandemic appearances of yellow rust are also known in wheat, as in 1955, 1957 and 1961 (Zadoks, 1961, 1965). In the case of wheat the physiological races varied per region in accordance with the dominant varieties of the regions. This fact supports the idea of the local origins of severe outbreaks. Weather effects similar over large areas and reigning over a period of a year or more are needed to produce yellow rust pandemics.

The information of the 'General Report' allows a generalisation on the long-term weather leading to the 1846 epidemic. The severe winter 1844/5, which had caused heavy frost damage to winter rye, delayed the spring growth of the fall-sown crop. I suppose that relatively much spring rye was sown, which ripens later than winter rye. The delays were not corrected, maybe even enhanced by a cool growing season, and the rye harvest was relatively late. I presume that yellow rust had an opportunity to multiply and spread unnoticed. The harvest was delayed by rains and storms.

More recent experience with wheat taught that such conditions may lead to many volunteer plants sprouting from shattered grain, not mentioned in any contemporaneous report, on which yellow rust can over-summer in abundance (Zadoks, 1961). The normal fall allowed timely sowing of the winter rye 1845/6. The 'green bridge' (Zadoks, 1984b: Figure 27.3) spanning the crop-free summer period was wide (much self-sown rye) and short (late harvest, timely seeding). Ample infection of the newly emerging rye crops was possible. The winter 1845/6 was exceptionally mild and allowed for more successive rust generations than usual. Every generation can produce a ten-fold multiplication of inoculum that may spread in all directions without meeting varietal barriers. Under these conditions[132] only a very hot and dry spring can stop the epidemic.

Rye is an obligate cross-fertiliser with pollination by wind. Most pollen will land near to its site of origin but some may be blown away over large distances, say hundreds of kilometres. European rye formed a meta-population with exchange of genes, including genes for susceptibility and resistance to yellow rust. The two major populations were winter rye and spring rye with different but overlapping flowering periods. Though landraces of rye must have existed in the 19th century I believe that genetic barriers between various sub-populations were not very effective against large-scale yellow rust epidemics. Epidemics may have had a suicidal effect by selecting for minor gene resistance in the rye generation following the epidemic, as Vanderplank (1968) suggested for maize and tropical maize rust in W Africa[133].

These epidemiological considerations are highly speculative. The sequence of weather types needed for over-summering, switchover, overwintering, and outburst should occur over a large area of continental Europe. A pandemic of yellow rust can be understood as the composite of many local epidemics, all driven by the same sequence of weather types over the period of about one year. For example, the year 1904 was favourable to yellow rust of wheat in several countries (Chapter 5), in Denmark also on rye (Rostrup, 1905: 355). The last large-scale epidemic of yellow rust on rye[134] known so far was in eastern Germany, Poland, Czechia, and Austria, 1916 (Chapter 5).

In Switzerland, 1957, I visited a breeder's field with much yellow rust on rye, amidst severely infected wheat fields. The lesions (stripes) were sharply delineated, chlorotic, poorly sporulating, i.e. of a moderately resistant type. Such an infection cannot be attributed to f.sp. *secalis*. The relatively susceptible rye selections (not commercial varieties) showed an infection 'mirrored' from the epidemic on wheat that was ascribed to the 'Probus race' of the f.sp. *tritici* (Zadoks, 1961).

2.4. Discussion

Early in my career, during the severe yellow rust epidemic on wheat, 1957, I was 'requested' (read ordered) to appear before the Counsellor for Plant Diseases and his number two at the National Agricultural Advisory Service in Wageningen, Netherlands. The conversation took place in a somewhat conspiratory atmosphere and astonishingly culminated in the question how to differentiate between brown rust and yellow rust. Where two pre-eminently able and well informed persons did not know that difference in 1957, differentiation must have been far more difficult in the mid 19th century. The gap between the two dates spans over a century of phytopathological research with great refinement of vocabulary and diagnostics.

Uredo rubigo vera, frequently mentioned as the 1846 rust on rye, is of a chimerical nature, yellow rust in spring, brown rust – sometimes – in mid-season, and yellow rust again towards ripening. Maybe, I am biased in favour of yellow rust, my first research subject. In my eyes, the evidence for a yellow rust epidemic on rye in 1846 is convincing with its early start, yellow dust, stripes, and with glumes, rachis, kernels and awns infected, as well as the inner sides of sheaths. The oft repeated observation of abortion of some grains and good filling of the remaining ones, a typical case of physiological compensation, records a new symptom, characteristic for yellow rust on rye and not seen on wheat.

Retrospective conclusions based on symptom descriptions not yet up-to-standards are risky, a high level of probability replacing absolute certainty. Accepting this reservation the conclusion is clear, yellow rust, *P. striiformis* f.sp. *secalis*. Rust on wheat, 1846, was mentioned incidentally; it could have been yellow rust but its appearance did not parallel the yellow rust on rye. Thus

I conclude that the f.sp. *secalis* was already in existence. The possibility that in some places brown rust dominated cannot be excluded but I don't have evidence in favour of this possibility. Mixed occurrence of the two rusts is thinkable, but in wheat it is observed only at low severities of either rust. At high severities one of the two rusts will dominate. The frequently mentioned rust on the heads, in combination with the colours yellow and orange, are major arguments to ascribe the damage on rye in 1846 to yellow rust.

Another argument is the typical damage to the grain. Severe rust on wheat causes shrivelling of the grain, at times with some abortion at the tip of the ear. This is what I have seen with yellow rust epidemics on wheat in the Netherlands. 'Chicken fodder', sneered the farmers[135]. Here I add the differential effect of yellow rust on rye kernels, some aborted, others more swollen than normal, without a fixed position of the aborted grains. This typical symptom, which I associate with the open flowering of rye, was not mentioned by Eriksson & Henning (1896) who – in their time – did not see a severe attack of yellow rust on rye. I did not find this symptom in other books.

A thorough, contemporaneous discussion of the 1846 rust epidemic on rye is not available[136]. My guess is that scientists were too busy with potato late blight (Chapter 3). The evidence brought together consists of numerous little scraps of information made meaningful by a few brief journal articles. Nonetheless, there is a fair degree of consistency in all these bits and pieces as to symptomatology and damage. In as far as I know, historians (Pirenne excepted) did not deal with the rye rust[137], even though the dearth of rye in 1846/7 was discussed extensively (Chapter 3).

The European dimension of the epidemic is still inadequately known; very severe in Belgium and the Netherlands, severe or worse in various German states, and present in N Switzerland and Denmark (then including Schleswig-Holstein). The epicentre of the epidemic seems to have been within the triangle NE Belgium, SE Netherlands, and the German Rhine Province. Did the epidemic have a wider reach? The answer is yes – Silesia and East Prussia, that is all of modern W Poland, apparently suffered from yellow rust, as did Brandenburg and Saxonia (Figure 2.13). I did not examine local newspapers and reports.

As an afterthought, an anomaly presses forward. Intense and prolonged drought in the East, and severe and continuous drought in the W Europe, reduced the rye yields (§3.2.6), whereas a disastrous yellow rust epidemic appeared in between. How is this possible? Where is the moisture needed by the rust to infect? The tentative explanation is twofold. (1) A phase difference may have occurred, the rust peaking earlier than the drought. (2) A few scattered records suggest lack of rain combined with high humidity. Clear skies and high humidity lead to heavy dew at night, providing the moisture (free water) allowing the rust to flourish[138].

The effect of the drought on rye can be equated approximately to the yield reduction of wheat that was in the order of twenty per cent (Tables 2.1 and 2.2; various authors) in Belgium, France and Germany. The excess loss beyond twenty per cent might be ascribed to yellow rust if we wish to ignore the effects of icy winds and night frost at florescence.

Yellow rust on rye is capricious, popping up here or there, and disappearing suddenly. The 1846 epidemic of yellow rust on rye was ephemeral. Its origin in 1845 is a matter of conjecture only. Its aftermath is unknown, a few scattered sentences offering no holdfast.

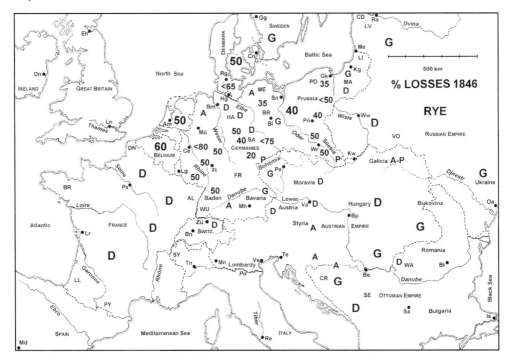

Figure 2.13. Tentative map of rye yields and losses in Continental Europe, 1846.
Numerals are losses in per cent (<80 means losses up to 80%) of the expected yields. The rationale for the expected yields was rarely indicated. One may think of the mean yield over the preceding 3 to 5 years.
A = average yield, D = considerable damage, G = good yield, P = poor yield. Large letters and numerals refer to national data, small ones to regional data. For details see Appendix 2.3. Losses may be due to drought (mainly in the E and W), cold winds, night frost, yellow rust, voles (= field mice), or combinations.
NB: Figure 2.11 provides geographic details. Figure 3.6b shows some losses in cereals, wheat and rye combined, whereas Figure 2.13 refers to rye only.

Well documented are the epidemic in Sweden, 1794[139], and the epidemics in Middle Europe, 1914 and 1916[140]. Sweden, late 19th century, and Denmark, early 20th century, had some irregular appearances[141]. Here, we add Continental Europe, 1846. So far, that's it. The 1916 epidemic is the last one on record for rye. The forma specialis *secalis* of the yellow rust seems to have disappeared completely.

On the political economy of plant disease epidemics

Table 2.2. Mean grain and straw losses in per cent for wheat and rye in the Belgian provinces, 'adjusted losses', and hectoliter weights of rye and wheat (calculated from data in Table 2.3; hectolitre weights from Anonymous, 1850b). 'Adjusted loss' of rye is rye loss in % minus wheat loss in %. 'Adjusted loss' is a proxy for loss due to yellow rust, total loss corrected for general weather effects.

Province	Grain loss in %		Adjusted grain loss in %	Straw loss in %		Adjusted straw loss in %	Hectoliter weight in kg	
	Wheat	Rye	Rye	Wheat	Rye	Rye	Rye	Wheat
Antwerp	11	56	45	12	22	10	72	79
Brabant	15	70	55	10	25	15	71	79
W Flanders	11	57	46	20	22	2	70	77
E Flanders	16	58	42	12	17	5	72	79
Hainaut	24	62	38	21	25	4	70	77
Liège	9	69	60	6	15	9	72	79
Limbourg	22	71	49	11	26	15	72	78
Luxembourg	25	64	39	24	24	0	70	76
Namur	19	46	27	16	17	1	70	76
Belgium	17	61	44	15	22	7	71	78

2.5. Conclusions

1. A yellow rust epidemic appeared on rye in 1846, with damages up to one half, occasionally even two thirds, of the expected yield.
2. The epidemic was ephemeral, appeared suddenly, and hardly left a trace.
3. The epidemic was severe in Belgium, the Netherlands, and Western Germany (Rhine Province, Saxony, Baden), and also in present Poland. The epidemic reached north to Denmark, east to the Baltic area and the Carpathian Mountains, south to Switzerland, and west to Belgium.
4. The rye crops suffered much from drought, in the East (present Poland) and in the West (France), also in parts of Scandinavia. In the East cold cutting winds in late May and early June were very damaging.
5. The yellow rust epidemic on rye has hardly been mentioned by historians; it was neglected by scientists, possibly because of the overwhelming impact of the 'potato murrain'.
6. The economic and social impact of the epidemic was considerable, since at the time rye was still the major food of the poor.

Appendix 2.1. Reasoning applied to identify the rust species, extracting the 'General Reports' for 1845 and 1846, supported by my own experience with the cereal rusts.

As this is a linguistic analysis I did not translate from Dutch to English. The font code is:
Headings and sub-headings in **bold italics**.
Literal quotations in **bold**.
Short, non-literal paraphrases between 'quotation marks'.
Comments in normal type.
The passages (in Dutch) are [numbered] and the summary remarks (in English) are [lettered].

Verslag 1845 - Milieu-factoren

Algemeen
[1] 'Door lange winter laat gewas'.
  Een laat gewas kan leiden tot een (iets) verlate oogst, en dat is gunstig voor het invangen en vermenigvuldigen van roest-inoculum in het voorjaar.
[2] 'Natte Augustus. Op 19 augustus woedde een zware zomerstorm (p. 8)'.
  Het lijkt onwaarschijnlijk dat op 19 augustus nog veel winterrogge op stam staat, maar bij zomerrogge kan dat misschien wel. Zomerstormen veroorzaken zaaduitval (bij tarwe soms tot ~1 ton per ha) en dus opslag. Zaadval uit garven aan het stuuk is alleszins denkbaar bij zware storm.
[3] 'Lange winter, laat voorjaar, vochtige zomer, gemis aan zonneschijn (p. 11)'.
  Een donkere, vochtige zomer met zomerregens vertraagt de afrijping en bevordert het schot, de kieming van uitgevallen zaad en het aanslaan van opslag-rogge, essentieel voor de overzomering van inoculum van ieder van de drie roestsoorten. Een late oogst gevolgd door een vroege zaai bevordert de oversprong van inoculum. Voor zwarte roest heeft de overzomering geen consequenties.

Limburg[142]
[4] 'Zaad met schot. Laat rijp, 13 dagen regen in de 1e helft van Augustus, dus schot, toppen van garven groen (p. 19)'.
  Het schot is hier niet relevant, maar de regen bevordert massale kieming van rogge-opslag (bij tarwe kan dat tot 1000 plantjes per m^2 gaan), die zelfs door goede grondbewerking niet totaal verwijderd kunnen worden.

Drenthe
[5] 'Regenschade'. Zie [4].

Verslag 1846 – symptomen, milieu-omstandigheden, schade

Winter 1845/6 (p. 12)

... een 'ongemeen zachten winter en een' buitengewoon warmen en droogen zomer, welke weêrsgesteldheid tot laat in de herfst heeft voortgeduurd. ... alle granen en vruchten vroegtijdig ingezameld en van de akker verwijderd ....

[6] Dus enkele extra generaties mogelijk gedurende de winter zoals ook voorafgaand aan gele-roestepidemieën op tarwe. Ook bruine roest kan van zachte winter profiteren (Chester, 1946; Zadoks and Leemans, 1984).

Limburg

... dat men nooit zulk eenen onvoordeligen Rogge-oogst ... heeft te betreuren gehad. 'Op veel plaatsen kwam het dorsloon er niet uit, dus oogst niet gedorst maar aan de schapen gevoerd (p. 27)'. **Zij** [de plant] **begon te kwijnen en in haar onderste bladeren geel en roestig te worden... De roest klom op in de schacht en tot in het onderste der aar, welker top ook nog door vorst werd aangetast; zij werd gedeeltelijk ja zelfs geheel geel, waardoor het korrelen volstrekt verhinderd werd, ...** (p. 27).

[7] **geel en roestig**
De gele kleur van het onderste blad duidt op gele roest.
[8] **klom op**
Het 'opklimmen' van de aantasting, tot in de aar, duidt op gele roest; bij bruine roest is het opklimmen afwezig of weinig opvallend en aar-aantasting afwezig.
[9] **aar ... gedeeltelijk ja zelfs geheel geel**
Aar-aantasting komt voor bij gele en zwarte roest, maar de kleur spreekt voor zich.

Noord-Brabant

..., toen men in het einde van April in de vroeg gezaaide Rogge een zeer verontrustend bederf bespeurde, namelijk het verdorren der onderste lissen of bladeren, die met eene klevige gele stof bezet werden, en zich meer en meer uitbreidende, binnen weinige dagen tot alle met Rogge bezaaide akkers overging, ... (p. 27). **Deze, tot dus verre, zelfs bij de oudste landlieden onbekende kwaal, die in den beginne algemeen voor honingdaauw gehouden werd, doch naderhand met den naam van roest werd bestempeld ...** (p. 28).

[10] **einde van April**
Zwarte roest kan worden uitgesloten op grond van het vroege verschijnen van de roest en van de verdere symptomen.
[11] **klevige gele stof**
Urediniosporen zijn geel en vormen bij sterke sporulatie een soort gele stoflaag op de bladeren. Het gele stof blijft aan handen, kleren en schoenen hangen of kleven en is moeilijk van de kleren te verwijderen. De kleur geel zal niet voor bruine roest gebruikt worden, dat ook kan kleven maar wel in mindere mate. Volgens Rapilly (1979) is gele roest zeer goed aangepast aan verspreiding door spattende druppels, hetgeen het 'kleven' zou kunnen verklaren.
[12] **zich meer en meer uitbreidende**
De term zou kunnen slaan op de semi-systemische groei van de gele roest, waarbij sporulerende lesies (vlekken op jong blad en strepen op oud blad) in de lengterichting van het blad uitgroeien. De term kan ook slaan op uitdijende haarden.
[13] **Onbekende kwaal**
Het is moeilijk voor te stellen dat bruine roest als een onbekende kwaal wordt gekenschetst. Bij gele roest is dat eerder denkbaar.

[14] **Honingdaauw**
In de 19ᵉ en 20ᵉ eeuw gangbare, aspecifieke term voor verschillende verschijnselen, waaronder gele roest[143] (Ponse, 1827). Kennelijk hebben landbouwkundig geïnteresseerden in 1845 de term al ingewisseld voor roest, daarbij de keus latend uit één van de drie roestsoorten op rogge (zie Box 2.4).

[15] **verdorren der onderste lissen**
Blijkbaar stierf het (onderste) blad geheel af, zoals bij een zware gele roest aantasting op tarwe.

[16] **'ging later in de aren over'**
Aantasting van aren is normaal voor zwarte roest en komt voor bij zware epidemieën van gele roest, maar niet bij bruine roest.

[17] **binnen weinige dagen tot alle ... akkers overging**
d.w.z. bij alle akkers voor de passant zichtbaar wordend binnen een korte tijdspanne. Bij gele roest vaak gezien, zou bij bruine roest kunnen maar minder opvallend zijn.

Noord Holland[144]

De Rogge heeft over 't algemeen in de maand Mei aan de roest geleden, die aan de bladeren een roodstofachtige kleur mededeelde, en zich van het onderste gedeelte van den halm tot aan de aren uitbreidde (p. 33). De Hr. Perk te Hilversum meldde: de beste rogge leed **het meest van de roestziekte, gaf weinig koorn, veel maar slecht stroo, over het algemeen geen half gewas** (8 i.p.v. 20 mud). **De kwaliteit goed en zeer zwaar, ...**

[18] **roodstofachtige kleur**
In mijn ervaring wordt de kleur van gele roest regelmatig als rossig aangeduid; het volledig verrueste en verdroogde onderste blad krijgt bij droog weer een roze tint. De kleuraanduiding pleit meer voor bruine dan voor gele roest. Merk op dat de kleur van bruine roest wel als rood maar nooit als geel wordt aangeduid. Von Schlechtendal noemde het sporenpoeder 'oranjerood', Braun schreef 'roodachtig geel' en 'van pommeransgeel tot fel menierood'. Beide auteurs hadden ongetwijfeld gele roest voor ogen.

[19] **van het onderste gedeelte van den halm ... uitbreidde**
Het van onder naar boven opstijgen van de aantasting is kenmerkend voor een vroege gele roest aantasting voortkomend uit ter plekke overwinterend (dus niet ingewaaid) inoculum. Bij bruine roest zou dit ook kunnen maar veel minder opvallend zijn. Bruine roest heeft een iets hogere temperatuur nodig dan gele roest en komt later op gang.

[20] **tot aan de aren**
Hoewel er niet staat 'tot in de aren' neem ik aan dat de waarnemer aantasting in de aren gezien heeft. Dit komt bij bruine roest niet voor.

[21] **kwaliteit goed en zeer zwaar**
Deze opmerking kan geïnterpreteerd worden als zeer beperkte vruchtzetting, waarbij de weinige fertiele vruchtbeginsels uitgroeiden tot goede zware korrels.

Utrecht

... zich vrij algemeen, doch het sterkst in de vroeg gezaaide [rogge], **roest op de halmbladen vertoonde, welk bederf zich ook tot aan de aren in de kafbladen en de korrel voortzette** (p. 34).

[22] **vroeg gezaaide** [rogge]

Vroeg gezaaide rogge kan door de gele roest al in het najaar geïnfecteerd zijn en zal dan in het voorjaar ook het eerst duidelijke symptomen laten zien, mede dank zij de semi-systemische groei van de roest in het overwinterende roggeblad. Vroeg gezaaide rogge kan ook bruine roest vertonen maar dat valt weinig op door de trage vermenigvuldiging in de winter en de afwezigheid van systemische groei.

[23] **tot aan de aren in de kafbladen**

Aantasting van de kafbladen of *glumae* is voor gele en zwarte roest kenmerkend. De gele roest sporuleert aan de binnenzijde, de zwarte roest vooral aan de buitenzijde.

[24] **en de korrel**

Als de *glumae* zijn aangetast door gele roest sporuleren zij aan de binnenzijde en bedekken de vruchtbeginsels met geel sporenpoeder. Sporen 'op de korrel' zijn vaak te zien bij een ernstige gele-roestepidemie. Bovendien zijn in zeldzame gevallen op het nog groene vruchtbeginsel kleine lesies te zien van niet-sporulerende gele roest.

Gelderland

De roestziekte, waardoor dezelve in het voorjaar is aangetast, heeft eerst voor een algemeen misgewas doen vrezen; ... 'Oogst tenslotte 1/3 van verwachte oogst. Volgens Wttewaal (1848)' **1°. de bladen door de *Uredo Rubigo vera* De Cand. 2°. de aren door dezelfde soort van roest, alsmede door de *Puccinia graminis* Pers.** (p. 36; conform Anonymous 1846c: 98).

[25] ***Uredo rubigo vera* De Cand**.

De Candolle's beschrijving dekt de gele roest zoals wij die nu kennen, zie Box 2.3.

[26] **de aren door dezelfde soort van roest**

Alweer, de aren werden aangetast en dat is kenmerkend voor gele en zwarte roest.

[27] ***Puccinia graminis* Pers.**

Nergens wordt een opmerking gemaakt over de voor zwarte roest kenmerkende halmaantasting, hetgeen tegen de identificatie pleit; de telia van de gele roest zijn ook zwart. De telia zijn bij zwarte roest groot en prominent, bij gele roest klein en weinig opvallend. Bij gele roest zien we vaak binnen één lesie gele uredinia alterneren met zwarte telia; soms neemt het zwart van de telia de overhand. De verwarring met zwarte roest ligt dan voor de hand (Anonymus, 1846c: 98).

Overijssel

... de bloei ontwikkelde zich niet, en er kwam ene gele stof aan de bladeren en den stengel, welke zich aan de aren mededeelde; de Rogge stoof niet of slechts weinig ... de ziekte, roest genaamd, vermeerderde sterk en deed zelfs de korrel aan (p. 39; conform Anonymous 1846c: 100).

[28] **ene gele stof aan de bladeren**

Het gele stof duidt opnieuw op de urediniosporen, die kennelijk in een dikke laag aanwezig waren.

[29] **en den stengel**

Dit is een nieuw element in de beschrijvingen. Het meest waarschijnlijk is dat de stengels dicht bij de bladaanhechting bestoven waren met sporen. Misschien moeten we denken aan uredinia op de halm, zoals die bij gerst nog wel eens voorkomen, meestal niet-sporulerend, en zelden bij tarwe. Halm-aantasting is kenmerkend voor zwarte roest, maar dan past het 'geel stof' niet.

[30] **de Rogge stoof niet**

Rogge is een obligate kruisbevruchter en windbestuiver. Bij bepaalde weersomstandigheden ziet met wolken sluifmeel over het gewas wervelen. Blijkbaar waren hier de planten zo zeer aangetast dat zij niet meer in staat waren stuifmeel te vormen of althans stuifmeel los te laten. Dit symptoom is voor mij nieuw.

[31] **en deed zelfs de korrel aan**

Waarschijnlijk afzetting van urediniosporen, gevormd in de kafjes, op de vruchtbeginsels.

Friesland

... waarbij in het voorjaar of het begin van den zomer zich nog de roest voegde. ... hier en daar zo erg, dat men het op eenen aanmerkelijken afstand kon zien. 'Bij Ternaard' **... vele velden geheel mislukt... Een eigenaardig rood uitslag op stengel en bladen scheen de hoofdoorzaak van die mislukking** (p. 40).

[32] **in het voorjaar**

De tijdsaanduiding sluit zwarte roest vrijwel uit.

[33] **op eenen aanmerkelijken afstand**

Kenmerkend voor ernstige lokale aantasting door gele roest, bv in haarden of vroeg gezaaide veldjes. Aantasting door bruine roest is op afstand moeilijk te zien.

[34] **Een eigenaardig rood uitslag**

Over de term 'rood' is hierboven al iets gezegd. 'Uitslag op de stengel' is kenmerkend voor zwarte roest, komt niet voor bij bruine roest, en treedt bij uitzondering op bij gele roest (zie boven). De opmerking kan ook duiden op afzetting van urediniosporen van gele of bruine roest.

Groningen[145]

... maar daarenboven ontdekte men zeer algemeen de roest (*Uredo rubigo*) op de bladen, die daardoor een ziekelijk-geel en roestkleurig aanzien verkregen. 'Vanaf 1 mei'. **De onderste bladeren waren op laatst van Mei reeds geheel verwelkt en ook de bovenste bladen door deze ziekte aangedaan. Bij het begin der bloeijing verspreidde dit kwaad zich ook op de aar, en wel hoofdzakelijk op de binnenzijde der *kafblaadjes* (*glumae*) en *klepjes* (*valvulae*), ja tastte ook enkele vruchtbeginsels zelve aan, en was als een geel poeder op de stempels zelve zichtbaar** (p. 40/1). 'Opbrengst 1/2 tot 1/3'. **... daar het roest bij enkele regens daarvan als afgewasschen werd, en den grond der Roggelanden als met een roestkleurig poeder bedekte.**

[35] *Uredo rubigo*

Een *nomen nudum*, een lege naam, die brand-roest betekent, zie Box 2.3.

[36] **een ziekelijk-geel en roestkleurig aanzien**

Duidt op gele roest, maar ook koude, stikstofgebrek en pathogenen kunnen verkleuring veroorzaken.

[37] 'vanaf 1 mei'.

Een begin-datum zegt meer over de betrokkenheid van de waarnemer dan over de ziekte. De datum is te vroeg voor zwarte roest.

[38] **onderste bladeren waren op laatst van Mei reeds geheel verwelkt**

Kenmerkend voor een vroege en zware gele-roestepidemie, bij bruine roest zeldzaam.

[39] **hoofdzakelijk op de binnenzijde der** *kafblaadjes*

Sporulatie op de binnenzijde der kafblaadjes is een onderscheidend kenmerk voor gele roest.

[40] **en** *klepjes*

Is hier sprake van actieve sporulatie of van bedekking door sporen afkomstig van kafjes?

[41] **tastte ook enkele vruchtbeginsels zelve aan**

Opnieuw de vraag: vertoonden de vruchtbeginsels lesies, hetgeen zeer wel mogelijk is bij vroege aantasting, of waren zij met sporenpoeder overdekt?

[42] **geel poeder op de** *stempels*

De stempels zijn kleverig om het stuifmeel op te vangen, maar zij kunnen dus ook urediniosporen 'vangen'. Voor mij nieuw symptoom. De vraag rijst of de aanwezigheid van een overmaat gele roest sporen op de stempels de bevruchting kan verhinderen.

[43] **het roest bij enkele regens ... afgewasschen**

Dit gebeurt inderdaad, vooral bij gele roest, waarvan de urediniosporen grote affiniteit tot water vertonen (Rapilly, 1979). Vraag is echter of de waarnemer het 'afwassen' zelf gezien heeft[146] of slechts daartoe concludeert.

[44] **den grond ... met een roestkleurig poeder bedekte**

Bij zware aantasting door gele roest ziet men vaak dat de grond onder het gewas geel of geelachtig kleurt door de 'sporenregen'. Dit gebeurt vooral als het waait, de halmbladeren tegen elkaar aan slaan en de sporen van het blad geschud worden. Ik denk dat de term 'regens' niet te letterlijk moet worden genomen.

Drenthe

Reeds vóór dat de Rogge in bloei kwam, vertoonde zich een roode roest op het loof ... dit ongemak nam gedurende den bloeitijd inzonderheid der Winter-Rogge sterk toe. Verwachtingen gunstig ware niet de algemeene ziekte, bekend onder den naam van rood of roest daarover gekomen (p. 44). 'Te Smilde halve opbrengst door rood of roest maar' **... hare hoedanigheid was uitmuntend, ...** (p. 45).

[45] **roode roest**

Aan de term 'roest' hoeft niet getwijfeld te worden, maar welke? De benaming van kleuren is verwarrend, zie boven.

[46] **rood of roest**

Hier is niet duidelijk of 'rood' een kleuraanduiding is of (tevens) een alternatieve naam voor roest, zoals in Overijssel[147].

[47] **hare hoedanigheid was uitmuntend**

Kennelijk heeft de roest zoveel abortie veroorzaakt dat de weinige fertiele vruchtbeginsels forse korrels konden produceren.

Gerst – Groningen

In de tweede helft van Maart en het begin van April had men hier en daar wel eenige roest (zoo het scheen dezelfde soort als die, welke op de Rogge later zoo algemeen is voorgekomen) opgemerkt, maar dit herstelde zich weder; ... (p. 53).

[48] **tweede helft van Maart**
Een goede waarnemer heeft al vroeg gekeken, in Maart. Vroege aantasting van wintergerst duidt op gele roest. Ook dwergroest begint al vroeg in het jaar maar die valt in Maart minder op.

[49] **dezelfde soort als die ... op de Rogge**
Dit is een betekenisvolle opmerking. Zo vroeg in het jaar is zwarte roest uitgesloten. Dwergroest van gerst (*P. hordei*) en bruine roest van rogge zijn nogal verschillend. Gele roest van gerst, rogge en tarwe zien er hetzelfde uit. De opmerking wijst sterk in de richting van gele roest.

Summary remarks

Symptoms *point to the identity of the rust at species level.*

[a] [23] [39] [40] Infection of glumes is characteristic of yellow and black rusts, even giving a name to yellow rust (*P. glumarum*), but does not occur with brown rust.
[b] [16] [19] [20] [26] infection of the ears is characteristic for yellow and black rusts.
[c] [11] [14] [28] (sticky) yellow dust (sometimes mentioned as honeydew) characterises yellow rust excluding the other two rusts.
[d] [12] 'extending' may indicate yellow rust, but the term is quite vague.
[e] [48] The similarity between the rusts on barley and rye strongly suggests yellow rust which occurs rather regularly on winter barley and, apparently, quite incidentally on rye.
[f] [24] [31] [41] Infection (if that is meant) of the kernel is rather specific for yellow rust but also rather rare. Spore powder on the kernel is seen with both yellow and black rust, probably more with yellow rust. Yellow rust sporulates on the inner side of the glumes, black rust usually on the outer side.
[g] [29] Rust (dust) on the stem may point to both yellow and black rust. The statement is too weak for black rust characterised by conspicuously large lesions on the stem.
[h] [42] Trapping of urediniospores by the sticky stigmas is a new symptom, non-specific, but indicative of the severity of the epidemic.
[i] [25] [27] [35] I think the Latin names are unreliable and non-discriminating (Box 2.3).

Field observations *cannot proof the identity of a rust but they may be indicative.*

[j] [38] Nearly complete killing of the lowest leaf is characteristic for a severe attack by yellow rust, but it is not impossible with brown rust.
[k] [37] An early date is characteristic for an early epidemic of yellow rust, less so of brown rust, and rules out black rust.
[l] [18] [34] [36] [45] [46] In my experience the term 'red' as colour indication is poorly discriminating, in contrast to the term 'yellow'. The term 'red' may also refer to brown rust and, even more so, to black rust.

[m] [13] The expression 'unknown ailment' agrees with my experience in wheat. Brown rust occurs nearly every season, yellow rust appears more as a surprise.

[n] [21] [47] In the obligately cross-fertilising rye fertilisation often fails due to a severe infection by yellow rust, but if it succeeds the resulting kernel is well filled. This phenomenon, which I have never seen in self-fertilising wheat, was mentioned explicitly by Braun (1846) and Treviranus (1846).

Environmental factors *cannot prove the identity of a rust, but they may add to the plausibility of the choice for one or another rust species.*

[o] [1] A late crop in the preceding year furthers the 'switchover' of the rust to the early- sown following crop.

[p] [2] to [5] A wet harvest period stimulates the over-summering of inoculum. Self-sown volunteer rye is essential but the 'General Reports' do not mention it. Points [1] to [5] are valid for yellow and brown rust, not for black rust.

[q] [22] Fields sown early in the fall trap inoculum of various pathogens among which yellow and brown rusts and hence promote the switchover of the rust from preceding to following crop.

[r] [6] A mild winter is essential for yellow (Zadoks, 1961: §32.2) and brown rust since it allows for more generations than usual and for a good quantity of inoculum to initiate the 'outburst'. Usually black rust does not normally overwinter in the uredinial stage in NW Europe.

[s] [10] [17] [32] [33] [47] Early infection, visible from a distance or at a field border, is characteristic for yellow rust, less so for brown rust.

[t] [43] [44] Often I saw a yellow coloured soil under a wheat crop severely infected by yellow rust, quite seldom I saw a brown coloured soil under a wheat crop severely infected by brown rust.

[u] [30] The lack of pollen shedding and dispersal of rye during a severe epidemic is a new symptom, but it is a non-specific symptom.

The most specific *symptoms* relate unambiguously to yellow rust, less specific symptoms do not exclude brown rust. Black rust can be excluded, primarily because the most important symptom, stem infection, has not been mentioned. *Field observations* clearly point to yellow rust but do no completely exclude brown rust. *Environmental factors* indicate yellow rust, analogous to the severe yellow rust epidemics on wheat in the period 1955 through 1961 (Zadoks, 1961). My conclusion is that the nation-wide epidemic on rye, 1846, was due yellow rust (see Utrecht and Groningen, also N Brabant), but that local outbreaks of brown rust or mixed occurrence of yellow and brown rusts cannot be excluded.

Appendix 2.2. Data from Belgium, 1846.

A detailed agricultural census was made in 1846 (Anonymous, 1850b). Rye covered 283,369 ha, wheat 233,452 ha, and potatoes 115,062 ha. Rye, the poor man's crop, was the most important arable crop. Most of the rye (4/5) was used to feed the Flemish population. At the time, rural Flanders impoverished, suffering from the decline of the cottage textile industry, whereas the Walloon (French) speaking part of Belgium, rapidly industrialising, suffered little from the harvest failures.

Provincial yield data were used to demonstrate the effects of the rust epidemic on rye. There are at least two – more or less – independent effects on yield, the general weather situation during the growing season and, in the case of rye, the putative effect of the rust.

The yield depression of rye was impressive, with a mean yield for Belgium of 7.25 hl/ha in 1846 versus a long-term average of 18.68 hl/ha (Table 2.3). A comparison with wheat provides more detail. The overall yield of wheat was 15.35 versus 18.41 hl/ha, a grain loss of 17 per cent and a straw loss of 15 per cent (Table 2.3). This loss cannot be due to a cold winter or a conspicuous disease or pest, but might be attributed to summer drought, as in France (Agulhon et al., 1976: 140; Jardin & Tudesq, 1973: 234; Le Roy Ladurie, 2004: 623). Projecting this weather effect on rye yields, the wheat losses in % are deducted from the rye losses in % and the resulting loss, the 'adjusted loss', is attributed to yellow rust. For Belgium as a whole this boils down to 44% grain loss and 7% straw loss.

The interesting point is the mean hl weight of 71 kg/hl, which is quite normal for rye (Table 2.3). Knowing that a severe rust epidemic occurred in rye, the combination of a large yield depression and, nonetheless, a good grain quality points indeed to yellow rust on rye. The limited 'adjusted

Table 2.3. Mean grain yields in hl/ha and straw yields in kg/ha for wheat and rye in Belgian provinces (Data from Anonymous, 1850b).

Province	Wheat yield				Rye yield			
	Grain (hl/ha)		Straw (kg/ha)		Grain (hl/ha)		Straw (kg/ha)	
	1846	Normal	1846	Normal	1846	Normal	1846	Normal
Antwerp	15.85	17.83	2731	3120	8.11	18.25	2708	3457
Brabant	13.94	16.45	2639	2946	4.97	16.64	2344	3127
W Flanders	17.86	20.14	2355	2947	9.92	21.23	2340	3017
E Flanders	16.82	20.10	2654	3002	9.09	21.64	2907	3518
Hainaut	15.00	19.67	2693	3411	7.74	20.25	2834	3765
Liège	15.55	17.04	2816	3005	5.47	17.85	2760	3235
Limbourg	13.40	17.15	2449	2739	4.47	15.39	1973	2675
Luxembourg	11.07	14.74	1658	2171	10.41	19.37	2218	2904
Namur	12.93	15.90	2512	3005	5.87	16.42	2389	2890
Belgium	15.35	18.41	2581	3047	7.25	18.68	2520	3223

straw loss' does not contradict this conclusion. For the explanation of the curious combination of good quality grain with large grain losses see text.

The northern provinces (Antwerp, Brabant, Limbourg and Liège) were relatively hard hit with grain losses from 45 to 60 per cent and straw losses of 9 to 15 per cent. The high straw losses, especially, suggest a relatively early attack. In the Netherlands the rye of the southern provinces (Limburg, Noord-Brabant), neighbouring Belgium, was also heavily damaged.

Appendix 2.3. Abstract from reviews by Löbe (1847) and Weber (1847) on grain production in Europe, with special attention for rye.

Fall-sown wheat and rye usually followed the same pattern of yield response to the year's weather (Chapter 4), but in exceptional years as in 1846 the responses diverged. Winter rye suffered more from drought, cold cutting winds, and night frost than winter wheat. In the North and the East of the Germanies the wheat was sometimes diseased, probably attacked by a rust, but in the western part the wheat was healthy and high-yielding, in contrast to winter-rye.

Retrospectives on the 1846 harvest by Weber and Löbe in the *Oekonomische Neuigkeiten* of 1847 provided a varied picture of the damages. The mild winter produced an early crop. In the East, Silesia and Poland, strong, cold NE winds in spring (late May, early June) damaged the crops, causing poor seed set in rye. Spring frosts, mentioned occasionally as a damaging agent, I interpret as night frosts to which rye is very sensitive during flowering and early seed set. In several areas the rye, usually sown on the poorer soils, suffered much from drought. In some areas patches with deaf ears were found that could have been due to early rust foci or, alternatively, to some root or foot rot, or to night frost in low lying patches, or to the cold winds impairing fertilisation. Rust was reported from Poland to Belgium and from Denmark to Switzerland. Which rust?

The reviewers did not describe the rust in detail but mentioned its occurrence on foliage, stems and ears. This probably excludes brown rust. The clue is in the remarks on quality. Black rust, if on the stems, hardly prevents seed set and causes all seeds to be light and of poor milling quality. In contrast, yellow rust on rye often prevents seed set and then the remaining kernels are heavy and of good quality. Several comments concur in low yields but good grain or milling quality. Key was a remark on Belgium, where the yield was poor but the quality good. We reasoned already that Belgian rye suffered from yellow rust exclusively or primarily. By extension I suggest that yellow rust was the main culprit in the Germanies.

Admittedly, the conclusion on yellow rust having been the dominant rust in the Germanies of 1846 is based on an indirect argument, but it is the best available; it does not exclude the occasional presence of brown or black rust.

The following is an abstract from the surveys by Weber (W) and Löbe (L), with page numbers added. Note the differences between the two reviewers, Weber reporting earlier and far more optimistically than Löbe. The remarks on good quality are written in *italics*.

Austrian Empire

Inner Austria (Styria, Carinthia, Kranjska)	Rye harvest about average. Wheat good (L123).
Lower Austria	Drought damage in cereals (W15). Winter cereals good (L123).
Upper Austria	Drought damage in cereals (W15).
Banat	(Present Romania) Cereals good (W15).
Bohemia	Cereals good, except Ore Mountains where rye failed (W15).
Croatia	Cereals good (W15). Rye early and satisfactory (L124).
Dalmatia	Cereals suffered from drought (L124).

Galicia	Rye poor, wheat quite good (W15).
	Rye moderate, wheat excellent (L123).
Hungary	Drought damage in cereals (W15).
	Winter cereals average (L123).
Moravia	Drought damage in cereals (W15).
Transsylvania	(Siebenbürgen, present Romania) Cereals good (L124).

The Germanies

Baden	Rye good (W14).
	Rye yields halved (L133).
Bavaria	Wheat excellent and rye fair. Familiar words such as yellow dust and honeydew in an incomprehensible text[148] (L133).
	Lower Bavaria – Bavaria's grain shed, excellent rye (W14).
	Lower Franconia – Cereals good but some night frost damage in fall-sown cereals (W14).
	Middle Franconia – Wheat very good (W14).
	Würtzburg – Rust in spring wheat (W14).
Brandenburg	Wheat good. Soon a disease appeared in rye, that in many areas, especially the lower ones, reduced the yield by one third or even by one half (L125).
Brunswick	Wheat good. Loss in rye 50 per cent (W14).
	Rye yielded less than half the normal yield due to rust, wheat poor, spring sown crops suffered badly from drought (L132).
Bremen	Rye average, wheat good (W14).
Galicia	Rye bad but wheat good (W15).
	Rye average (L123)
Hanover	Rust reduced rye yields considerably (L132).
Hesse	Rye poor (L133).
	Rye yields halved by rust (L133).
Marken	(area around Berlin) – Rye good (W6).
Mecklenburg	Rye yield medium (L132).
	Mecklenburg-Schwerin – Cereals good (W14).
Oldenburg	Rye yields moderate (L132).
Ore Mountains	Rye with beautiful straw but few grains (W14).
Pommerania	Cereals good (W12).
	Rye suffered from fungal disease (G: *Befallen*). Harvest was poor. Rye fields had large patches with wholly white, deaf ears (L125).
Poznań	Cold nights with cutting winds (May/June). Rye yields often halved. Wheat yields good (W12).
	Cereals ripened prematurely because of the heat (L125/6).
Prussia	East Prussia – Grain yields good (W12).
	West Prussia – Rye good, some fields with poor yields excepted (W12).
Rhine Province	Rust in rye and wheat, straw and grain yields not abundant (W13).
	Newspaper quote 'grain yields good, but not rye that suffered in places from *rust, mildew and honeydew ...*' (W13).
	Rye yields down to one fifth (L125).
Saxony	Grain yields average, but rye often unsatisfactory (W13).
	Rye with good average yield. In many places yield unsatisfactory, but *with nice flour content* that compensates the low threshing (W13).
	Wheat good but smutty (L125). Rye average because of rust (L125).

	In most fields rye suffered from rust, grain loss about one third, but *good quality* compensates yield loss (L131).
	According to a governmental publication the overall losses of rye in 1846 were 22 percent[149].
	Saxon Duchies – Wheat excellent, rye in some areas yields one quarter of normal (L132).
Schleswig-Holstein	Yields less than half the expected value, but *grains heavy and rich in flour* (L132). Holstein – In the S losses in rye up to two thirds (W14).
Silesia	Cereals satisfactory, spring cereals moderate due to drought, rye had not enough kernels (W5).
	Winter cereals suffered from cold winds and night frost. Wheat coloured red by rust. Later rye became rusted too (L125).
	The rust coloured all wheat red (brown leaf rust, or yellow rust?) and later rye became infected too. 'Under these conditions the ears remained either completely deaf, or very incomplete. *The remaining kernels, where they could be formed, are very rich in flour*'. Yields approximately halved. Wheat satisfactory. Spring-sown cereals poor (L125).
	Upper Silesia – Grain yields moderate to bad. From mid May icy NE winds, early rye frozen. Winter rye, delayed by low temperatures, became rusty. Rye with rust that destroyed large areas (W5).
	Middle Silesia – Rye often 1/3 to ½ less kernels but *these were of good quality and they give very nice flour* (G: *die jedoch sehr schönes Mehl geben*, W6).
	Lower Silesia - Rye sometimes good (W6).
	Silesian mountains – Rye and wheat good (W6).
	Upper Lusatia - (Oberlausitz) Cereals good (W6).
Westfalia	Münsterland – 'Rye gives little promise of a good harvest as it seems to be affected by rust, its red dust and moist rotting' '... *the kernels swelled and were well developed and healthy; ...*' Yield in retrospect good. Note: 'In places *honeydew* befell the rye, as probably meant by 'rotting'' (W13).
	Rye yield meagre due to rust (L125).
Würtemberg	Cereals average (L133).
Belgium	Rye good (W15).
	Rye harvest nearly failed but *rye quality good*. Wheat yields fair. Spring sown cereals suffered from drought (L132).
Denmark	Wheat good, rye and spring sown crops lost 50 per cent due to unfavourable weather (L134).
Poland	(East of river Wisła, under Russian rule) Rye not so good (W15).
	Bad harvests. Rye suffered from spring frosts and fungal disease (G: *Befallen*); large patches with white deaf ears (L135).
The Netherlands	Rye gave normal yield on three quarters of the fields (W15).
	'The rust much reduced the rye yield' (L132).

3. The potato murrain on the European Continent and the revolutions of 1848

The tale of the Irish Famine, 1845-1847, following the outbreak of potato late blight, has been told repeatedly. The parallel story of the 'Continental Famine', 1845-1847, has not yet been told. The Continental Famine was caused by poor harvests of potatoes, due to the same late blight, and of grain, due to frost, drought, rust, voles, inopportune rains, floods and hailstorms. The Continental Famine was enhanced by hoarding, speculation, and poor governance. Hunger was followed by infectious disease. The demographic effects are difficult to disentangle. The number of excess deaths due to the Continental Famine cannot yet be determined with any precision, but clearly it approaches that of the Irish Famine. The harvest failures of 1845 and 1846 and the resulting famine came on top of rural pauperisation and urban discontent, and thus contributed to the revolutions of 1848 on the European Continent. The statement 'an epidemic of potato late blight caused an epidemic of revolutions' is largely exaggerated but it contains a grain of truth.

Ce végétal est bien, comme il l'a dit, le plus beau présent que le Nouveau Monde ait fait à l'Ancien.
This vegetal is really, as he said it, the nicest present that the New World has given to the Old.

> Antoine Parmentier, quoted by Augustin Sageret in the lemma *'pomme de terre'* (potato) in the Encyclopédie méthodique (Tessier *et al.*, 1818: 713).

3.1. The continental Famine

Phytophthora infestans[150] caused an epidemic disease in potato, then indicated as the 'Potato Murrain', that destroyed the potato crop in Ireland and triggered the 'Great hunger' (Woodham-Smith, 1962) of 1845-1847, also called the 'Irish Famine'. The infection was imported from the Americas[151] into Belgium on new potato breeding material and it began already to spread in 1844 (Bourke & Lamb, 1993: 5, 35). The year 1845 was a blight year but, from a meteorological point of view, not an extreme one[152]. Winds dispersed the inoculum in all directions over Europe, also to the NW, so that England and later Ireland were affected (Bourke, 1964). The Irish crop was exceptionally promising in 1845 until the blight killed the crop, sometimes within a week, a real catastrophe. In the following year, 1846, the blight was disastrous again, due to the weather, the late planting of the crop, and – supposedly – the tremendous amount of inoculum coming from culled potatoes left about hither and thither.

The Irish were poor and they had little else to eat than potatoes. Under the 'potato economy' an Irish labourer ate ~5.4 kg of potatoes a day[153], spread equally over three meals, as long as potatoes were available, roughly from November 1st to May 1st. Then came the summer dip, with hunger and disease. Failure of the potato meant famine, and a famine was accompanied by its usual complement of infectious diseases, intestinal infections, dysentery, typhoid, typhus, and tuberculosis[154]. We cannot differentiate the numbers killed by hunger from those killed by disease[155]. The results of the Irish Famine are well known, with ~600,000 excess deaths and an emigration of ~1,300,000 people over the five-year period 1846/51[156].

On the European continent the rural poor were, maybe, slightly better off than in Ireland. Their diet consisted largely of potatoes *and* rye. In 1845 and 1846 their subsistence was threatened by harvest failures in several major food crops, potatoes foremost. A famine followed, here called the 'Continental Famine'. The course of events on the Continent showed some similarities with that in Ireland but its impact on society differed. In Ireland a breakdown of traditional society took place with mass emigration but without political renewal, whereas on the Continent a renewal of societies followed the revolutions of 1848. This paper[156a] discusses the relationship of the phytopathological events in 1845-1846 and the political events in 1848.

Each section begins with some general comments, applicable to much of Continental Europe (Figure 2.12), and continues to discuss the events in different countries, the Netherlands (Figure 2.7) foremost.

3.2. Potato late blight and other discomforts

3.2.1. Potato cultivation and potato use

Between 1750 and 1850 the potato gradually gained ground upon the cereals, primarily rye (e.g. Bieleman, 1987: 543), though with large phase differences between countries and regions (Oliemans, 1988; Reader, 2008). Yields increased considerably between 1812 and 1845 (Van Zanden, 1985: 166). Providing one to two times more calories per hectare than cereals, and being a sturdy crop with regular yields, the potato became the staff of life for the poor, at least for the rural poor. Many urban poor, however, had bad housing without cooking facilities, and thus depended on bread only, rye bread of course (Jardin & Tudesq, 1973: 234).

Many potato varieties were in use, early, mid-late, and late. The late varieties should have a good keeping quality since storage facilities were minimal in comparison with today. Keeping quality had already become a problem in Ireland when the preferred potato variety changed from 'Apple' over 'Cup' to 'Lumper' in a period of about one hundred years (Bourke, 1993 p33). The average per capita consumption of potatoes in Ireland was ~800 kg/y, in the Netherlands[157] ~210 kg/y. The last figure seems representative for Continental Europe where calorie intake was supplemented by rye for the poor and wheat for the rich. In times of scarcity the calorie intake may vary considerable (Sen, 1981; Drèze & Sen, 1989). Heldring (1845) listed a weekly budget of a poor rural worker's family in the Netherlands (Table 3.1). No other data for the years around 1845 were found.

The area under potato increased and part of the new produce was destined for industrial processing. Potato flour (farina) was a new product. Distilleries produced malt wine, alcohol, and vinegar. The vinegar served to savour the potatoes. The alcohol was used in the spirit burner. Malt wine was the raw material of brandy. Use and abuse of brandy were considerable. The average (!) male in the rural town of Goes (Zeeland, NL) consumed more than half a litre

Table 3.1. Weekly budget of an average family of rural poor, consisting of three adults and three children, according to Heldring (1845: 28). He compared the new prices, after the blight of 1845, with the old prices before the blight. Prices are in Dutch cents. The casual labourer usually earned 5 cents per hour, in times of unemployment (as in 1845) even less. This budget dates from the fall of 1845; prices peaked in the spring of 1847.

Item	Old price	New price, fall 1845	Increase in per cent
Bread	90	140	56
Fat	35	35	
Butter	50	50	
Meal and buttermilk	50	75	50
Coffee	10	11	10
Treacle	5	5	
Oil	12	12	
Soap	5	5	
Cloth, thread, buttons	10	10	
Tobacco	5	5	
Salt	10	10	
Milk	7	7	
Pepper	3	3	
Vinegar	5	5	
Fuel	50	50	
Rents	50	50	
Clothes	100	100	
Pots, pans, ironwork	15	15	
Potatoes, pulp or groats	50	140	180
Totals	542	708	31

of brandy per week. Pastor O.G. Heldring, deeply concerned with the condition of the poor, thought that the potato contained a 'principle of sin' because it was also used to produce *jenever*, the Dutch version of gin[158]. He saw the punishing hand of God in the potato blight (Heldring, 1845). In the province of Brabant[159], however, people were made to believe that the blight was due to the introduction of the 'polka', a dance of Bohemian origin popular in Vienna since 1829 (Van Oirschot *et al.*, 1985: 4).

3.2.2. Potato diseases

The major potato disease during the late 18th and early 19th century was the 'curl', known since 1747 in Germany and 1764 (or 1751) in England. Today curl is attributed to a virus disease, possibly potato virus Y (Salaman, 1949). Control was difficult but not impossible. The most radical solution, though with a temporary effect only, was the selection of new varieties from seedlings (Van Bavegem, 1782), a current practice during the 19th century[160]. 'Renewal of seed', i.e. importing seed potatoes from less affected areas[161], was practiced as a control measure since ~1800 in Belgium and England[162]. A more subtle method of control was a form of 'green lifting'[163], mentioned in Friesland (NL), 1807 and 1809, and in Scotland[164].

In the early 1840s an epidemic of 'dry rot' occurred Europe-wide, due to a *Fusarium* fungus (Von Martius, 1842)[165]. In 1844, Flemish authorities, wanting to control the diseases 'curl' and 'dry rot', planted potato tubers imported from North America on an experimental farm at Cureghem in West Flanders[166]. After a while, some of the resulting plants showed curious brown flecks. Some of the harvested potatoes rotted in storage. This is the putative onset of the potato late blight epidemics in Europe (Bourke & Lamb, 1993: 5) that caused the Great Hunger 1845/47 in Ireland and the Black Years 1845/9 in the Netherlands[167]. The story of the 'potato murrain', a designation introduced by Berkeley (1846), in Ireland has been told over and over again.

The reports on 1846 rather frequently mention poor tuber formation, the number of tubers being sometimes down by fifty per cent or more (e.g. Hlubek, 1847: 69). The contemporaneous explanation was the response of the potato plants to excessive heat leading to long roots and few tubers[168].

3.2.3. The late blight epidemics of 1844, 1845 and 1846

1844. Late blight first appeared in 1844 in Belgium. Bourson (1845) impartially reviewed opinions on the causes of the 'evil' (F: *le mal*). His report on the 1844 events in Belgium is blurred by the then reigning confusion between 'curl', dry rot and wet rot, some savants contending that the three were – in fact – one. Several provincial authorities, reporting a good 1844 harvest, acknowledged that the potato quality was often poor in low places, an observation not incompatible with the occurrence of late blight[169]. Bourson remained quite vague on blight in Belgium, 1844, only quoting Kickx and Mareska, professors at the University of Gent, who reported on the disease 'that had wrought its ravages in secret' (F: *elle avait ... exercé sourdement ses ravages*).

Desmazières, botanist and mycologist, had seen the disease in 1844, in the department 'Nord', in N France[170]. He gave a detailed description of the symptoms, mainly on the variety 'blanche tardive' [= late white], and of the associated fungus that he called *Botrytis fallax*. The losses around Lille were 'considerable'[171]. The disease was seen frequently in SE England (near Folkestone in Kent), where it arrived too late to seriously harm the foliage but early enough to cause severe tuber infection leading to great storage losses (Mickle, 1845). Vis[172] described the loss of the tasty variety 'Beaulieu' in Zeeland (SW Netherlands) in 1844 due to a disease that we now recognise as late blight.

1845. The winter 1844/5 was extremely cold and frost may have killed culled potatoes thus eliminating inoculum. It also killed much fall-sown cereals and winter colza (rapeseed)[173] in the Low Countries and in M and N Europe. As a consequence farmers planted more potatoes than usual in the Netherlands, Belgium, and N Germany[174]. Seed potatoes and groundkeepers[175] may have carried inoculum through the winter. No relevant information is available. In 1845 the blight burst out in W Flanders (Bourson, 1845: 2720) and then spread over Europe as a first order 'focal epidemic'[176], wonderfully illustrated by Bourke (1964) on the basis of hundreds of original documents.

Bourke & Lamb (1993) analysed the seasonal weather for 1845. In most of NW Europe the weather, 'drab, cloudy and cold' (*ibid.*: 41), was but moderately favourable to infection. Winds could disperse inoculum from the focal centre in all directions. Extreme susceptibility of the potato crops rather than exceptional weather conditions determined the severity of the epidemic (Vanderplank, 1968: 153). The blight, beginning in West Flanders and severe already in June, hit the potatoes early and hard in the Belgian sea polders. Münter (1846: 8) noted a steep gradient in the wetness of 1845, from England (wet, cold, bad harvest), to NE Europe (warm, dry).

In the Netherlands the alarm bells tinkled everywhere in August. The early potatoes had been lifted safely, but the mid-late and late varieties became seriously diseased, sometimes up to total loss. The sandy soils, less conducive to late blight, could still produce a crop, though even in the peat-sand area of Drenthe the yield fell from ~210 hl/ha in 1844 to ~90 hl/ha in 1845, a loss of ~57 per cent (Bieleman, 1987: 650).

1846. The mild winter of 1845/6 must have allowed the ample overwintering of inoculum in culled potatoes, potato piles, groundkeepers, and seed potatoes, all over Continental Europe, but this was poorly documented[177]. In Belgium, Quételet (1846) observed flowering plants[178] throughout the winter. Late winter and early spring were unusually mild.

The summer of 1846, with its anti-cyclonal weather pattern over the continent, was, according to Bourke & Lamb (1993: 49), dry and hot, and according to several contemporaries exceptionally dry and hot. Drought damaged many crops over large areas. The rye crops suffered much in France and in present Poland, and in several hill districts. In Lower Silesia (present W Poland) the potato harvest failed by drought, not by late blight. Similarly, the early potatoes in Bavaria – escaping the blight – could not form decent tubers by lack of water.

The statement by Bourke & Lamb that the potato crop 'suffered more from drought than from blight' is strong but not far from the mark. Löbe (1847) and Weber (1847) reviewed the information on the 1846 harvest in most of Continental Europe. The early potatoes generally suffered little from late blight and excessive heat, and - usually – gave a fair yield. The late potatoes, however, suffered much from heat and drought, as tuber formation was poor in many areas. In Belgium and Schleswig-Holstein the number of tubers per stool was low. In the Rhine Province 'late potatoes yielded only few tubers' (G: *Spätkartoffeln liefern nur wenig Knollen*). The same complaint, 'very few tubers' (G: *sehr wenig Knollen*), came from Saxony. In Poznań the potatoes failed, because 'so far often only long roots were formed in stead of tubers' (G: ... *statt der Knollen häufig nur lange Wurzeln bis dahin angesetzt haben*). In Poland potato produced about one third of the expected number of tubers and, occasionally, no tubers at all but only long roots (Weber, 1847: 12). Many tubers stopped growing until the rains came and then produced secondary growth. I did not (yet) find the complaint about poor tuber formation in the Netherlands. Everywhere, loss by late blight came on top of loss by poor tuber formation.

The German daily *Allgemeine Zeitung* of 2 August 1846 wrote 'Everywhere the continuing heat melts the ice on the mountains. The summit of the Mont Blanc today presents naked rocks, during many years the ice there did not disappear. In consequence several rivers burst their banks, such as the Rhône, which flooded some 1000 *Juchart* (~300 ha) of cultivated fields'[179]. Floods – destroying the crops – were reported from i.a. the Rhine in Lichtenstein and the Vistula (P: *Wisła*, G: *Weichsel*) in present Poland. Clear skies might have led to severe night frosts damaging the young potato crops and the flowering rye crops but the available information is not satisfactory. Nightly dew was probably frequent. In many areas yellow stripe rust (*Puccinia striiformis*) of rye took its toll (Chapter 2).

Thunderstorms interrupted the fine weather and hailstorms could be very destructive. One such hailstorm in Bavaria destroyed the food base of 15,000 people[180]. The additive effect of these storms must have been considerable thus fuelling scarcity and distress. The thunderstorms in July provided the wetness needed by late blight to explode. Exploding it did, and around the end of July messages came in from Copenhagen and Gdańsk (G: *Dantzig*) in the North to the Savoy in the South, from Ireland in the West to Bavaria, and – later – Ukraine, in the East.

3.2.4. Countries

Agriculture knows good years, when everything works together in favour of the crops, 'normal' years, and bad years, when wrong goes what can go wrong. 1846 was a bad year. Among the mishaps were drought and excessive heat, yellow rust on rye – the staple for many Europeans –, voles, hail storms, and floods. The potato murrain was the last of the mishaps, picking up speed in August and becoming very destructive all over continental Europe. The 1846 data suggest an E-W gradient, a weather gradient, in contrast to the W-E gradient of 1845, a dispersal gradient.

The Netherlands. The subjective 'feel' of a season has been captured in a single figure[181], the 'frost index' for the winter and the 'summer index' for the summer (Figures 3.1 and 3.2). Temperature (monthly mean of daily temperatures) and precipitation (in mm per month) were plotted as the deviations from their 30-years' (1831-1860) means (Figures 3.3 and 3.4). The winter 1844/5 was bitterly cold[182] with monthly mean temperatures ~5 °C below normal in December, February, and March. The winter of 1845/6 was warm with monthly means of 2-4 °C above normal from November through March. The summer of 1846 was hot[183], the second hottest of the century. The winter 1846/7 was very cold again with monthly means of at least 3 °C below normal in December and January. Monthly precipitations were about normal in 1845 and 1846, with the exception of a high precipitation in December, 1845.

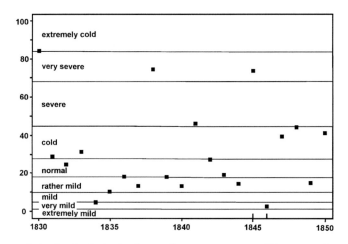

Figure 3.1. The Frost Index for the Netherlands over the years 1830-1850 (IJnsen, 1981).
The value of e.g. 1840 should be read as the value for the winter 1839/40. The winter 1844/5 was very cold, the winter 1845/6 was exceptionally mild. Horizontal – years. Vertical – Frost Index (ranging from 0 to 100).

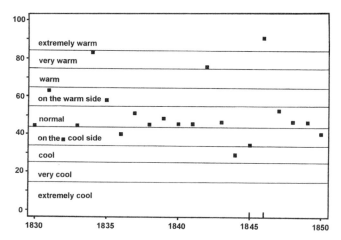

Figure 3.2. The Summer Index for the Netherlands over the years 1830-1850 (IJnsen, 1976).
The blight year 1846 was exceptionally warm. Horizontal – years. Vertical – Summer Index (ranging from 0
to 100).

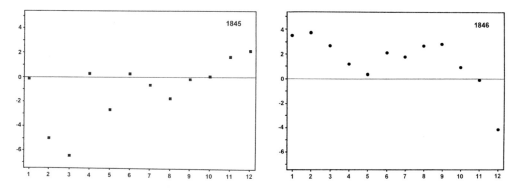

Figure 3.3. Monthly means of daily temperatures, De Bilt, the Netherlands in 1845 and 1846, plotted as
deviations from the 30-years' averages (1831-1860). Horizontal – months of the year beginning with January.
Vertical – Deviation in °C of the mean monthly temperature of 1845 from its long term mean (mean monthly
temperature of 1845 minus mean monthly temperature averaged over 1831-1860). Original data from General
Reports.

In the course of August, 1845, people realised that a catastrophe drew near. The Dutch
government reacted promptly. On 14 September 1845 the King signed an Order of Council,
proposed 9 September, to withdraw import duties on food commodities. Later, 16 November,
the Minister of Finance Van Hall declared in Parliament 'All other artificial measures lie, in my
conviction, beyond the circle of duties of the Government'[184]. Dutch potato yield in 1845 was
only ~0.27 million tonnes (~3.9 million hl; General Report, 1846), nearly 60 per cent less than
needed.

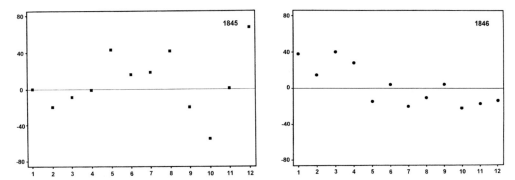

Figure 3.4. Monthly means of precipitation in De Bilt, the Netherlands, 1845 and 1846, plotted as deviations from the 30-years averages (1831-1860). Monthly precipitation in mm. Original data from General Reports.

In 1846 the area under potatoes decreased in some provinces, as in the province of Noord-Holland (Van Ewyck, 1847: 43). In the province of Friesland, the blight in 1846 was worse than in 1845 (Wumkes, 1935: 225). In the province of Groningen the blight caused serious damage again. In 1846 the total potato production of the Netherlands was ~0.4 million tonnes (~5.8 million hl; General Report, 1847). The 1846 deficit was clearly due to late blight but differed in structure from the 1845 deficit[185].

In the Netherlands the blight epidemics had several agricultural consequences. (1) The area under potato was reduced, by 35 per cent in Groningen, by 75 per cent in the Beijerlanden, south of Rotterdam, and by up to 100 per cent in the hard hit area of the Bommelerwaard[186]. The old levels were regained with a delay of some 20 years. Potato was replaced by leguminous crops for food and mangolds for feed. (2) The potato area shifted somewhat to the sandy soils of Drenthe and Brabant, where the micro-climate was less conducive to blight[187]. (3) Early potatoes received more attention as they often escaped the blight. The produce was readily exported to the Dutch towns and abroad, to England foremost. (4) Several potato processing plants had to close down. In Groningen, only two distilleries out of sixteen survived the 1846 crisis (Priester, 1991: 372). Dekadal average yields in hl/ha decreased from 185 (1831-1840) to 126 (1841-1850), recovered slowly to 148 (1851/60), to reach the nearly normal level of 181 (1861-1870) with considerable delay (Priester, 1991: 350).

Austria[188]. In 1845, blight arrived in the Austrian Crownlands but the amount of damage is uncertain. Probably there was just enough blight to infest the seed potatoes. The field crops of Austrian Silesia (present SW Poland) yielded poorly, about 25 per cent less than in 1844. Grain planting in the fall of 1844 and the spring of 1845 had been difficult or impossible due to excess wetness of the soil. Many potatoes set in the spring did not emerge. The grain and potato harvests 1845 suffered of bad weather (Nitsch, 1846).

In 1846 the potato harvest was ruined in Silesia and Galicia (both in the S of present Poland), Bohemia and Moravia (present Czech Republic), and the Bukowina (present Ukraine and Rumania). In Bohemia and the Bukowina losses were ≥75 per cent (Sandgruber, 1978: 64). The disease was impressive in Styria and serious in Carinthia and Carniola [Kranjska][189]. Tuber formation was poor in Inner Austria leading to over 30 per cent loss (Weber, 1847: 5).

In 1847 the blight struck again causing an empire-wide loss of 27 percent[190].

Belgium. The first outbreak op potato late blight in Europe occurred in Belgium (§3.2.3). Unfortunately, I could find few details.

In 1845 the blight was early and severe. Tubers rotted in the soil and the 1845 harvest was negligible, ~9 per cent of the expected harvest[191]. In the Flemish parish of Lippeloo people flocked in from all sides to invoke Saint Anthony on behalf of the potato. Coins piled up in the choir and had to be carried away by the basket (Lindemans, 1952: 189). The population was hard hit because many grain fields had been lost by freezing in the winter 1844/5.

The potato failed again in 1846. Actual mean yield in 1846 was 133 in stead of an expected 200 hl/ha, or a country wide loss of ~33 per cent. The worst hit province was E Flanders with a yield of 106 in stead of the expected yield of 200 hl/ha, a loss of ~47 per cent (Anonymous, 1850b).

The potato harvest of 1847 was not good but no useful data were found.

Denmark. The Duchies Schleswig and Holstein[192] suffered great losses in 1845 (Kyrre, 1913: 68). About 50 tonnes of potato bought in Denmark, seemingly healthy, were shipped to Hamburg. Upon arrival they were rotten and had to be thrown into the sea (Morren, 1845a: 30).

In 1846 the blight was more severe than in the preceding year (Löbe, 1847: 134), also in the duchy of Holstein. Export contracts had to be cancelled in 1846.

Estonia (then under Russian rule). In 1846 several potato crops suddenly turned black and wilted after flowering, without night frost. Healthy and diseased tubers were found in one and the same hill. Where late blight had not infested the crops, the potato harvest, with very small tubers because of the heat and drought during summer, was of no importance (Anonymous, 1847d).

France. In 1844 the blight appeared in NE France, bordering Belgium (§3.2.3). Decaisne (1846) quoted Desmazières, an able mycologist from Lille, who in 1844 already had seen an outbreak of '*Botrytis fallax*' on the variety *Blanche tardive* (= 'late white') in N France and had described the symptoms.

In 1845, French growers, seeing the majority of the tubers rotted, were so depressed that they often abandoned their potato fields (Roze, 1898: 290) and even 'believed to be obliged to abandon the cultivation of that precious tuber' forever (Mangin, 1914: 93). Locally the 1845 losses were very high, about 90 per cent in the district Lille, in the NE, ~45 per cent in the Savoy, and up to 100 per cent in some districts of SW France[193]. Country-wide, the potato yields in both 1845 and 1846 were 78 in stead of the expected 103 million hl, a loss of 22 per cent (Agulhon *et al.*, 1976: 140). In 1847 yields were normal again.

The Germanies[194]. The harvest of 1844 was good. In 1845 the weather had a gradient from very wet in W to hot and dry in NE Germany. Hence the losses by blight were up to 100 per cent in the W, where the consumption quality suffered too, and negligible in the E (Münter, 1846: 6/12). Münter's map clearly shows the blight gradient. In the W the tubers had been infected in the field, but in the E the disease was sometimes only found in the potato storage piles[195] (Münter, 1846: 28/9). Overall, the potato harvest was 50 to 75 per cent below expectation in 1845 (Wehler, 1987: 643; see Table 2.1). In some areas, as in Silesia, the wet harvest weather in combination with a mild winter caused great losses by warming and rotting in storage piles (Schulz, 1846).

In 1846 losses were estimated at ~47 per cent. Potato suffered from drought and late blight[196]. Complaints about deficient tuber formation were general (Löbe, 1847: 132). In several areas drought did more damage than late blight. Many crops, too bad to be harvested, had to be ploughed in.

Even the 1847 potato harvest was bad (33 per cent below expectation), a promising crop being destroyed by blight when the weather turned foul at mid-summer. In Upper Silesia the 1847 loss approached 100 percent[197].

Norway (then under Swedish rule). The murrain reached Norway late in 1845 (Bourke, 1964)[198]. In SW Norway the blight was even worse in 1846 and the potatoes rotted again (Lamb, 1995: 22; Löbe, 1847: 134).

Poland (modern E Poland, then under Russian rule). The potatoes failed in 1846, as did the grain. 'The distress is thus indescribably great'[199].

Sweden. The 1845 epidemic spread N up to the area of Uppsala, with great destruction in the southernmost tip of the country (Eriksson, 1884). In 1846 the blight in Östergötland was more severe than in 1845 (Löbe, 1846: 134).

Switzerland[200] with its rapid population growth was hit hard by a famine in 1845/6 triggered by the potato late blight (Gruner, 1968: 43, 47), 'a warning from heaven' (G: *ein Mahnmal vom Himmel*). Bonjean (1846: 116) provided interesting detail about this 'place of disaster'. In the Rhine Valley two thirds of the harvest was lost and a lack of setting potatoes resulted. In 1847 the Swiss author Jeremias Gotthelf published a tear-jerking novel 'Cathy the Grandmother' describing the disastrous effects of the 1845 and 1846 blight on her little household, a 'potato mini-farm' (G: *Kartoffelzwergwirtschaft*).

3.2.5. Interaction between harmful agents

A difficult and neglected subject is the interaction between harmful agents (Van der Wal & Zadoks, 1971). Around 1900 it was common knowledge that 'curl' enhanced the effect of potato late blight. Quanjer (1913: 62) observed that the rather blight resistant cultivar Paul Krüger could be severely attacked by late blight when diseased by the curl. J.C. Dorst, who later became a well-known plant breeder, made a pertinent remark to this effect, and he referred to an English publication of 1872, making an equally pertinent statement[201]. It seems as if this knowledge has been lost. We have, unfortunately, no way to establish the relative impacts of the two diseases and the importance of the interaction. Accepting the statements at face value, the interaction must have enhanced the damages caused by the two diseases.

3.2.6. Other crop damages

The year 1845. The 1845 season was a poor season for cereals being either too cold, or too wet, or too dry. The winter 1844/5 had been long and severe and fall-sown cereals had died by frost all over the Continent. In the Netherlands, Belgium, France, N and M Germany winter cereals were often replaced by potatoes (sic!).

The Netherlands. Winter wheat was poor and weedy, spring wheat did well. The rye harvest was variable, failing in Limburg. In some other areas about one third got lost due to wet harvest conditions (General Report, 1845).

Austria. Disastrous floods occurred in Galicia (present Poland, river Vistuła), Lombardy, Bohemia[202], and Hungary ruining the grain harvests (Macartney, 1968: 312). Rye yields were generally low (Abel, 1974: 367). In Austrian Silesia the wheat was severely rusted and yielded far less than in 1844 (Nitsch, 1846).

Belgium. Fields with winter wheat and winter colza had been destroyed (Reynebeau, 2005: 64). For that reason more potatoes had been planted than usual in Flanders.

France. The French grain harvest was moderate, 113 million hl where 120 million hl was expected, a loss of ~6 per cent (Agulhon *et al.*, 1976: 140). In places, less than 50 per cent of the expected rye yield was harvested. It is not clear what caused the loss, possibly winter damage. In Brittany the winter grain was frozen and many cattle died from cold (Guin, 1982: 117).

The Germanies. Winter killing of cereals was widespread in N and M Germany (Abel, 1974: 366). In Silesia rye suffered from voles, hail, and floods (Münter, 1846: 12). The Silesian grain harvest was less than 60 per cent of the expected yield (Schulz, 1846). The wheat harvest was ~1/3 below normal in the Rhine Province and poor in W Prussia[203]. In Mecklenburg and Silesia the wheat suffered badly from an unidentified rust; rye harvests were poor[204] because of voles, floods, and hail. The 1845 autumn in Upper Silesia was so wet that winter rye could be sown in two thirds of the area, leading to scarcities in 1846, as in Austrian Silesia. In parts of Masuria the rye did not produce its own seed[205].

Poland. The grain harvest failed.

The year 1846. The winter was mild, the grain harvest was early, and harvest conditions were good. Mild winters are favourable to cereal rusts, especially those without (active) perfect stage, allowing the production of a few extra generations so that the new growing season begins with abundant inoculum (Hogg *et al.*, 1969; Zadoks, 1961). The summer of 1846 was extremely hot over most of Europe and disastrous thunderstorms occurred in various areas, in the alpine area accompanied by summer floods. Other summer floods were caused by the rapid melting of alpine glaciers.

The Netherlands. A fitful winter followed by a dry summer often leads to an outburst of voles[206]. This happened in 1846 (Hooijer, 1847) when voles and drought together caused a national wheat loss of about thirty per cent (Van der Heiden, 2001). Wheat was not only damaged by voles but also by floods, birds, and (in the NE) rust. In Friesland, where fewer potatoes had been planted than usual, vole damage to cereals endangered food security (Jansma & Schoor, 1987: 59).

Rye became severely rusted by yellow stripe rust and the rye harvest of 1846 generally failed (Chapter 2). The damage was the more unfortunate since the poor, deprived from potatoes, now had to pay dearly for their rye bread (General Report, 1846). The fall-sown crop of 1846/7 was, again, damaged by voles that continued eating young sprouts in the fall and winter, sometimes under the snow cover.

Austria. In the Austrian realm the cereal and hay harvests failed again, leading to great shortages all over Central Europe and to hunger typhus in Styria, Bohemia and Silesia. A thunderstorm caused great damage in Bohemia[207] on July 6th. Prices went up and in several provinces people suffered real famine (Springer, 1865: 136).

Belgium. Flanders (W Belgium) saw its rye harvest reduced by 60 per cent, from 18 to 7 hl/ha (Lamberty, 1949: 136). On the poor soils of the 'Kempen' (N Belgium) loss was about 50 per cent (Vanhaute, 1992). Two historians[208] ascribed the loss to rust, now identified as yellow stripe rust (Chapter 2).

France. In France[209], the prolonged dry and hot spring of 1846 and the heavy summer rains caused important losses in cereals. The grain harvest amounted to 91 in stead of an expected 120 million hl, a loss of ~24 per cent.

The Germanies. Grain harvests were miserable in the Kingdom of Prussia with a rye yield of 43 per cent and a wheat yield of 23 per cent below the mean yield 1841/5 (Chapter 2) See Table 2.1; Wehler, (1987: 643); several authors reported severe rust in wheat as in Brandenburg and West Prussia[210]. A traveller reported the terrific effect of drought on rye in the NE of Prussia[211]. Thunderstorms with hail and floods hit the country, W Prussia in particular; in Bavaria hail storms completely destroyed the crops of 15,000 people[212]. Grain prices were excessively high because of the dearth. The Poznań area (present W Poland) was struck by famine in July, 1846. In some areas the country folks, without potatoes, bread and salt, had to eat weeds boiled in water[213].

Switzerland. The alpine valleys were threatened by floods. The princedom of Lichtenstein was flooded by the river Rhine[214].

The year 1847 was a year of recovery in most European countries, but not in parts of Austria, where the grain harvests failed for the third time in succession, and in Silesia, primarily due to floods[215]. Famine was general in some eastern areas, as in the south of modern Poland. In W Europe grain prices plummeted in 1847 when a good harvest announced itself.

3.2.7. Scientific interest

Turner (2006) aptly described the scientific efforts after the famine, 1845 to 2000. Here we focus on scientific efforts during the Continental Famine. Scientists in various countries immediately zoomed in on the blight problem in a nearly feverish activity, producing a flurry of lectures and publications. The fierce fight between the 'fungalists' and their opponents[216], the anti-fungalists, was depicted by Large (1940), Ordish (1976: 132ff), Peterson (1995), and Semal (1995) with a lovely touch of drama; it does not need to be discussed here. Many explanations of the disease were proposed (Box 3.1). In the end, the fungalists won.

Box 3.1. Contemporary explanations of the 'potato murrain'.

Numerous explanations of the potato disease were proposed, supported by more or less serious argumentation. A contemporaneous overview was found in Rüter (1950 p414ff) quoting 1845 reports from provincial governors to the Dutch Home Secretary. Here follows a brief summary of some opinions from the European Continent.

Atmospheric influences. A special weather constellation was a current, non-committal explanation proposed by several committees (the Brussels Committee Commissions; *ex* Berkeley, 1846: 23; Committee, 1845) and by Dumortier (1845: 291); Harting (1846: 56) 'cosmic' [= meteorological] influences (fogs); Schacht (1856: 10), and others.

Excess nitrogen. An excessive multiplication of nitrogenous substances in the parenchyma in the above and below ground parts of the potato plant is the source of the recent epidemic (Unger, 1847: 313; and others).

Heat. Enklaar (1846: 268/9); Mauz (1845: 29), a country physician, experimented extensively with potatoes. He concluded that the summer heat in July, 1844, was the cause of the disease. The potatoes might have been sensitised by a hidden agent in 1845, in view of the poor flowering of the potatoes in that year.

Wetness. Many authors attributed the disease to excessive wetness, more or less in accordance with the 18th century humoral theory of disease. Professor Blume (1845) from Leiden University used the term 'hydropisy' (= dropsy) and Rupprecht (1847) 'telosepsis epidemica', in agreement with Unger's (1847) 'stagnated plant sap'. Harting (1846: 56) quoted 'telluric' influences (clay soil, wet, manured) enhancing disease.

Inner Life. The learned Director Gebel presented his opinion to the Prussian Academy, Session 53: 30 January 1847. He ascribed the disease to 'the weakened inner life of the potato'[217]. A similar argument is the 'sunken life energy of the plants' (Nitsch, 1846).

Dry rot. Several scientists confused late blight (wet rot) with dry rot, due to a *Fusarium*, that had caused epidemics just before 1845 (Von Martius, 1842), stating that the new disease was just another form of the old one (e.g. Bergsma, 1845: 12, 15; Dumortier, 1845: 291; Harting, 1846: 269).

Cryptogamic entity. In 1844 Desmazières described a fungus, which he named *Botrytis fallax*, as the cause of the hitherto unknown leaf flecks on potato (*ex* Decaisne, 1846). The Belgian lady-mycologist Marie-Anne Libert published a brief symptom description on 19 August 1845, and proposed the name *Botrytis vastatrix* (*ex* Semal, 1995). More or less simultaneous and independent were statements to this effect by Morren in Belgium and Montagne in France, both in 1845 (*ex* Berkeley, 1846: 36).

French observers compared the progress of the disease with an 'epidemic', meaning a large-scale outbreak of disease as among humans or cattle. The conclusion that late blight was dispersed by the wind was rarely drawn explicitly in the 1840s. Field observations induced Van Peyma and others to conclude to aerial dispersal of the disease[218]. The modernist Morren (1845a: 13,15, 1845c, Figure 3.5) had a clear understanding of the spores, produced in enormous numbers, travelling far, 'swirling' through the air in order to infect fresh foliage[219].

A conservative view, however, held 'that this disease is not an epidemic one, that means is not generated by a disease substance present in the air, but, when it transfers itself from a healthy to a diseased, that happens only by direct contact, as with fruits' (Schirm, 1846). The editor added a note 'Science surely has not yet demonstrated a miasmatic epidemic – without

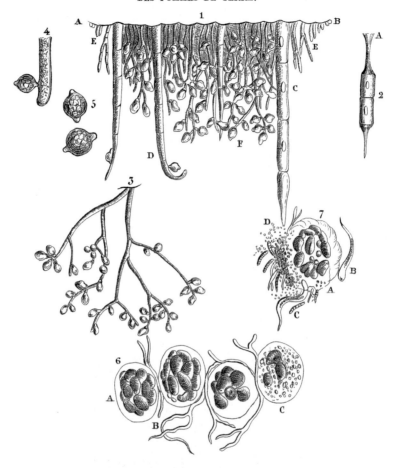

Figure 3.5. Note on Botrytis the destroyer or the potato fungus (Morren, 1845c p287). The drawing is probably by Morren himself; the paper is dated 24 September 1845. The explanation of Figure 2.1, given in the text, is rendered here in modern terms.

1 A..B - Lower leaf surface, C, D - leaf hair, E - Young mycelium, F – 'forest' of mycelial threads.
2 Leaf hair, top and bottom dry.
3 The large number of spores (sporidies) formed.
4 Fungal thread with spore still attached.
5 Spores of Botrytis with two papillae (mamelons).
6 Plant cells with starch grains, A - healthy cell, B – germ tubes of zoospores, branched, intracellular, C – cell content decomposing, filled with yellow granules that might be the reproductive bodies of the Botrytis.
7 Last stage of rotting, A - cell with degenerating grains, B – vibrions [bacteria?] or worms [nematodes?], C..D – fusarium spores (fusispores).

On the political economy of plant disease epidemics

contact – in the plant kingdom', miasmatic to be read as wind-borne, an echo of the 18th century miasmatic theory of disease persisting far into the 19th century[220].

In 1845, governments responded quickly to the emergency and ordered both seed potatoes and potato seeds from abroad for experimentation. They also requested their scientists to look into the matter and suggest methods of control. Two lines of research were developed. One line was to save whatever usable was left in the diseased tubers, by dry storage, producing dry potato chips, extracting the undamaged starch, or feeding the diseased lots to either cattle or distilleries. Diseased tubers, were they poisonous? The other line was to control or prevent disease by appropriate storage methods for healthy seed potatoes, selection of more resistant potato lines, correct dosage of manure, and – in the long run – selection of new varieties from seedlings (Rapport, 1845). Several scientists observed varietal differences in susceptibility to the blight[221].

The Groningen Committee published sensible recommendations on 16 September (Committee, 1845). 'The foliage of the diseased potatoes, that has no value, be burnt in the field and all useless and rotten potatoes be destroyed, so that as little as possible of the sickly crop remains on the field'[222]. If the disease reappears in the next season the affected foliage should be mowed or burnt. Spraying chalk water or, better, diluted sulphuric acid was recommended. Seed potatoes should be taken from slightly affected fields on sandy soils, and only from healthy stools. Tubers should be planted in rows and plants be earthed up.

Diseased foliage should be cut and burnt, diseased tubers burnt or fed to farm animals[223]. Chemical control was recommended by treating soil, seed tubers or plants with diluted sulphuric acid or by chalking them[224]. Morren was near to the mark with his recommendation to treat soil or tubers (not plants!) with a mixture containing copper sulphate. He observed the progress of the disease from foliage to tuber and he demonstrated experimentally the infection from tuber to tuber. Morren[225] (1845b) noted that the metallic fumes of the zinc factories had completely protected the potato crop[226], but the observation did not lead to further action.

The Netherlands. Dutch scientists showed a keen interest in the new disease. Bergsma (1845: 9, 15, 30) inspected 'several hundreds of hectares'. He wrote[227] 'Not rarely one saw a potato field completely changing in appearance within a few days and one perceived, especially in the evening, an unbearable stench[228] that was dispersed over a considerable distance'. 'The observation that the disease in its spread often has followed the direction of the wind becomes the more probable as some potatoes, growing behind hedges or trees, remained unaffected and only later contracted the disease'. He believed in the fungal origin of disease and became convinced 'that the *curl, rust* and *cancerous disease* do not differ from each other'. His recommendations were those generally given to control the potato dry rot. Moleschott & Von Baumhauer (1845: 3), studying the new disease in their spare time, found what we call now the sporophores and the oval spores of *P. infestans* (ibid.: 8) but could not convince themselves that the new fungus was the cause of the disease. They gave the usual recommendations. Harting (1846) looked into the botanical side of the blight[229]. Before the 1845 epidemic had run its full course Vissering (1845) already discussed the socio-economic implications of the potato murrain. The (liberalised) commodity market should do its work and, if the market failed, charity should step in.

At the request of the government Vrolik (1845) published his *ad hoc* field observations. The Dutch government collected potato seed from several origins and invited knowledgeable gentleman-farmers to grow the seeds. Mr H.C. Van Hall gave a sensible opinion on the 1846 tests[230], 'That the experiments, though, as regards the disease, not having answered objectives, however, by producing new sorts on very different soils … not have been without importance, in many respects instructive and promising not entirely unfavourable results for the future' (Vrolik *et al.*, 1846). The results were of little avail but we need not doubt the selection, in time, of new and (slightly) more blight-resistant varieties[231]. Vanderplank (1968) argued that the pre-scientific selection of varieties with at least some resistance was reasonably successful in the 19th century. Grootegoed (1853) regretted the loss of several tasty varieties[232] and the necessity to eat varieties grown as cattle fodder. English potatoes had to be planted, less susceptible to blight.

Belgium. In Belgium Bourson (1845) probed into the origin of the potato murrain. Du Mortier (1845) published his field observations with a good symptomatology. He had seen the *Botrytis* which he considered to be the consequence of a disease, more specifically the wet form of dry rot epidemic during the early forties. Morren (1845a, b), in contrast, was a convinced fungalist. The palm of honour I award to a lady mycologist, Marie-Anne Libert, one of the first to identify a fungus as the cause of the disease, *Botrytis vastatrix* (Semal, 1995)[233].

Denmark. The Danish authorities were quite alarmed in 1845. They sent out enquiries about possible protective measures and requested some of their embassies to find information and seeds. Three professors of the Polytechnic School in Copenhagen began research on disease control, without obtaining useful results; the causal agent was not yet known to them (Kyrre, 1913).

France. The blight was the subject of a lively discussion e.g. in the *Comptes Rendus de l'Académie* (Proceedings of the Academy), Paris, see its volume of 1845 and the contemporary list of references in Decaisne (1846). Decaisne was a good microscopist who saw the same sporophores and spores as Montagne. Nonetheless he decided that the '*Botrytis*' was but the consequence of the disease.

The Germanies. The Royal Prussian Academy of Sciences requested the botanist Münter (1846) to study the blight. He carefully collected national and international information on the 1845 outbreak and on the response of various authorities to the emergency. A scientific committee, meeting in Nürnberg on 22 and 24 September, 1845, produced sensible recommendations for damage control[234]. Many authors discussed the blight (see e.g. the 1846 volume of the *Botanische Zeitung*). The focus was often on the processing of infested lots of potatoes to utilise the remains as potato meal, alcohol, or cattle fodder. The detailed examination of the sporulation process, with elegant drawings of the fungus sprouting forth from the stomata, brought Unger (1847) near to the fungalist theory, though he was unable to sacrifice his ideas on the exanthematic[235] origin of plant disease.

Sweden. Several scientists took an interest in the new disease. The Royal Agricultural Academy asked professor Wahlberg (1847) to report on the disease. The famous mycologist Fries participated in the debate from the beginning and, interestingly, was against the fungal theory of disease causation (Eriksson, 1884: 6ff).

3.2.8. The quality of the information

The quality of the information on potato yield losses tends to decrease with the distance from the Netherlands, the bibliographic centre of this study. The information provided by historians passed through several filters, with the risk of distortion. Fortunately, the Official Newspaper of the Dutch Government (D: *Staatscourant*) of 1845 contained an unusually high number of scraps on agriculture, obviously important to a nation obtaining its wealth from international trade and shipping (Appendix 3.1). This contemporaneous information confirmed the data provided in the more general history books quoted in the present chapter.

Similarly, the German daily newspaper *Allgemeine Zeitung*, July 1846, contained scattered pieces of useful information, again providing contemporary confirmation of the historians' data. Its July message was simple, the potato crops were just fine. In contrast, the July issues of the Dutch daily *Nieuwe Rotterdamsche Courant*, 1846, reported that the blight was popping up, with local appearances scattered all over the Netherlands. The newspaper was even accused of needless alarmism but the editors took a firm stand stating that their task was to provide good information.

An enquiry by the National Economics Board of the Kingdom Prussia, late in 1846, arrived at a national loss figure of ~47 per cent (Anonymous, 1847a; See Table 2.1), the best available estimate based on experts' opinions.

Figure 3.6 sketchily relates the losses in potato and cereals, as far as reasonably known, to the areas involved, for 1845 and 1846.

3.3. Social consequences

3.3.1. General setting

In the Netherlands the position of the rural poor, and of the landless labourers foremost, had deteriorated considerably between the mid 18th and mid 19th century[236]. Wages were sadly low, 'the labourer earns too little to live, too much to die' (Brugmans, 1929: 135). Most labourers of the large-scale farms in Zeeland (NL) lived in abject poverty (Priester, 1998; Hoogerhuis, 2003). Farm workers often rented small plots, sometimes at usurious prices, to grow potatoes with and for their family[237]. Thus they could carry their family through the winter when they were without work and income, and sometimes they even could feed a hog. Potatoes are relative poor in protein but the protein is of a good nutritional quality. Potatoes contain much vitamin C so that the health situation of the underprivileged improved (Van der Heide, 2001), with e.g. scurvy disappearing[238].

The blight brought destruction of the potato, hunger and famine to the already destitute country dwellers. Whereas the situation in 1845/6 was bad but not unbearable (Bourke, 1983; Hooijer, 1847) it became really disastrous in the winter of 1846/7 (Hooijer, 1847). The Dutch situation is thought to be fairly representative of Continental Europe at large.

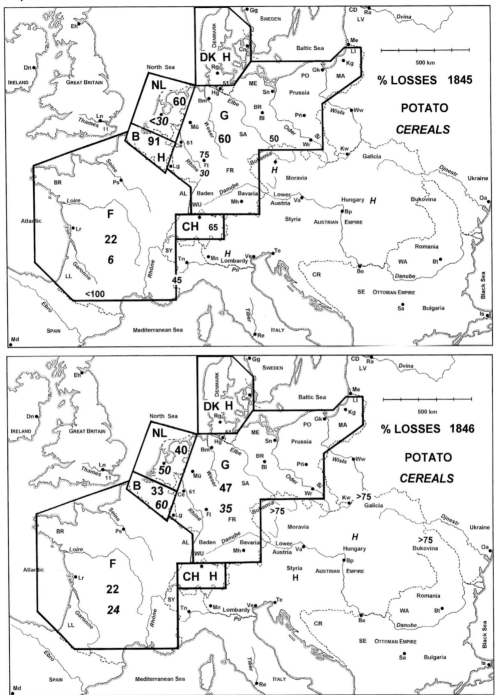

Figure 3.6. Sketchy maps indicating the relationship between loss figures, as far as reasonably known (see text), and areas involved for 1845 and 1846. The larger numerals represent national losses in per cent of potato (roman) and cereal (italics) yields. The smaller numerals refer to regional losses. H (= high) stands for economically and socially important losses. B = Belgium, CH = Switzerland, DK = Denmark, F = France, G = the Germanies (primarily Prussia), NL = Netherlands. See legend of Figure 2.11 for other abbreviations.

On the political economy of plant disease epidemics

3.3.2. Demographic effects

The period 1750-1850 was characterised by a rapid population growth in many European countries[239], including Belgium, England, France, Ireland, Prussia and Switzerland. During the half-century following the Napoleonic period the annual growth rate of the Dutch population oscillated around 1 per cent, most of the time (Figure 3.7). Mouths had to eat and hands had to work. The masses could be fed because potato growing became popular, potatoes producing one to two times more calories per ha than rye[240]. Rye remained the number one in the popular diet and potatoes came second, occasionally even first. Work for the hands declined on the countryside because the cottage industry, spinning and weaving, was gradually replaced by textile factories. The competition by modern industry caused a loss of labour opportunity and of purchasing power to both cottiers and rural craftsmen.

Among the demographic effects of the famine were an increase of the death rate, especially of children and elderly, and a decrease of the marriage and birth rates[241] (Table 3.2). In some areas as in Austria, Flanders, and the Netherlands (e.g. Zeeland) the population decreased in absolute numbers. Areas that had recently seen a rapid population increase, such as the Dutch clay soils, suffered most. As in Ireland, it is impossible to differentiate between deaths by hunger and by disease, since epidemics were the normal corollary of famines (Dyson & Ó Gráda, 2002; Sen, 1981).

The number of excess deaths due to the famine and its corollary diseases is heavily debated. The numbers for Ireland, 1845/9, vary from 500,000 to two or three millions[242]. Critical demographers now seem to prefer the lower number of 500 to 600 thousands. The numbers of excess deaths in Continental Europe can only be guessed at, examining population numbers and death rates. The numbers provided here (Table 3.3) are calculated as excess deaths in 1846, 1847, and 1848 by subtracting the average number of deaths over 1841-1845 from the number of deaths in the given years. The result must be considered with reservation as it needs critical

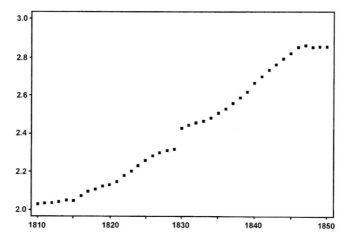

Figure 3.7. Population growth in the Netherlands from 1810 to 1850. Growth begins ~1815 and continues up to 1847, followed by a slight but telling decline in 1848. The sudden rise in 1830 cannot be explained. Data from Hofstee (1978), excluding the province of Limburg.

Table 3.2. Birth rate, death rate, and population growth rate in six countries, data in per mil, averaged over 5-year periods. The increase in death rate and the decrease in birth rate from 1841/45 to 1846/50 is seen in all six countries. Among the causes of excess mortality are hunger, typhus, and cholera. Dead-borne children are excluded or ranked under mortality (Switzerland). Migration is disregarded.

Country	Birth rate		Death rate		Population growth rate[7]	
	1841-1845	1846-1850	1841-1845	1846-1850	1841-1845	1846-1850
Austria[1]	13.0	11.2	29.6	35.8	9.7	1.5
Belgium[2]	32.1	28.5	23.2	25.2	8.9	3.3
France[3]	28.1	26.7	22.7	23.9	6.8	2.2
The Netherlands[4]	36.3	33.6	25.7	30.2	13.3	13.8
The Germanies[5]	36.8	35.6	26.1	27.4	1.0	0.2
Switzerland[6]	32.3	29.4	24.3	23.0	8.0	6.4

[1] Bolognese-Leuchtenmuller (1978); Sandgruber (1978); emigration negligible.
[2] André & Pereira-Roque (1974).
[3] Braudel & Labrousse (1976); usually immigration surplus. Population growth rate data in Armengaud (1971: 31) slightly different; the period 1846-1851 showed a net emigration of 276,000 persons (*ibid.*: 32).
[4] Hofstee (1978); high growth rate because of immigration.
[5] Estimated from Bolognese-Leuchtenmuller (1978); Sandgruber (1978).
[6] Bickel (1947).
[7] Emigration was of little importance in the numerical sense but emigration to other continents was of great significance from a political point of view. Transmigration within Europe was considerable.

examination by professional demographers. The data in Table 3.3 include deaths by hunger ad corollary diseases (mainly typhus). The year 1849 was not included because an epidemic of cholera spread over Europe killing many people, especially the undernourished, and so the total may have been under-estimated.

The Netherlands. Excess deaths due to hunger have been estimated at ⩾53,000[243] on a population of ~three millions. Among the corollary diseases were typhoid, dysentery, cholera, and malaria (Jansma & Schroor, 1987: 59). Symptoms of dysentery are recognisable in the description of the wretched poor dying in the Bommelerwaard, a district in the Netherlands where people were as dependent on potatoes as the Irish (Box 3.2). Malaria was common in the coastal areas of the Netherlands. The brackish waters along the coast provided a good breeding habitat for malaria mosquitos[244]. Malaria was a nuisance rather than a killer but it killed those debilitated by lack of food[245]. At the time, the death rate in Zeeland could be over 4 per cent, children, elderly and itinerant labourers, all with little immunity, dying first. In some Zeeland communities infant and child death peaked[246] in the famine years, especially in the extremely hot year 1846 with lack of food and fresh water (Hoogerhuis, 2003: 273).

For some of the diseases the temporal relationship with the famine years is more evident than the causal one. The hot and dry summer of 1946 created horrible conditions in crowded cities such as Amsterdam, leading to a high death rate[247]. During the cold winter of 1847/8 influenza struck the population (Terlouw, 1971: 288). In 1848/9 a wave of cholera swept over Europe,

Table 3.3. Tentative estimates of excess mortality due to potato late blight and its corollary diseases, 1846-1848. For the calculation see §3.3.2. I did not consider 1849 in order to avoid most of the mortality due to cholera. Cholera swept over Europe 1848-1849 and lingered on for years. Starved patients will succumb earlier to cholera than well-fed persons, but it is questionable to consider cholera as a 'corollary' disease.

Country	Population size 1846 (thousands)	Excess mortality 1846-1848 (thousands)	Annual excess mortality 1846-1848 (per thousand)
Austria[1]	17,902	371	7
Belgium[2]	3,933	30	3
France[3]	36,097	109	1
The Netherlands[4]	3,054	26	3
The Germanies[5]	33,197	213	2
Switzerland[6]	2,330	0	0
Total	96,513	749	3

[1] Estimated from Bolognese-Leuchtenmuller (1978).
[2] Hofstee (1978), present borders.
[3] Estimated from Dupâquier (1988), NB excess mortality mainly because of cholera epidemic.
[4] Hofstee (1978), mortality excluding ~22,000 cholera victims.
[5] Estimated from Bolognese-Leuchtenmuller (1978).
[6] Based on Bickel (1947: 135 and 148).

Box 3.2. Quotations from the Reverend C. Hooijer (1847).

The parson C. Hooijer made a passionate plea in favour of the deprived in the Bommelerwaard, a river clay area West of Arnhem (NL) where typical 'Irish' conditions prevailed among which (1) potatoes were the dominant crop with ~26 per cent of the arable area, in some communities even ~60 per cent (Bieleman, 1992: 152), (2) the poor depended on potatoes for food, (3) labourers rented small potato plots at high, often usurious rents. I translated some selected remarks:

1. During the winter 1845 farmers gave potatoes to the poor. These ate only potatoes, cooked at noon, porridge in the evening, baked in the morning.
2. 1846 ... large groups of beggars, who could no longer be fed by the citizens ...
3. Because of the dearth relief paid only one quarter of the food needed. People consumed things green from the land, boiled in water with some barley meal. Dung-hills were searched for eatables.
4. ... chew willow bark to silence hunger. ... steal and cook cats and sick goats ...
5. ... people wither away ... hovels with in a corner nearly naked boys who, in their third year, cannot yet walk.
6. ... dry mother breasts with crying sucklings ...
7. ... in dark huts spectre-like persons who, in the most disgusting filth, with the most dismal indifference laid down to die as beasts ...

causing ~22,000 deaths in the Netherlands[248], and reached Ireland in 1849. Malnutrition may have sensitised cholera patients but here we consider the cholera epidemic as independent from the 1846/8 famine. Poor nutrition had the usual effect on the growth of youngsters. The percentage of undersized conscripts (less than 1.57 m tall) was 14.2 in 1821 and 19.8 over 1850-1861 (Van Zanden, 1991: 22).

Austria. Annual population growth rates were -0.69 for 1847/8 and -1.28 for 1848/9 (Bolognese-Leuchtenmuller, 1978: Tables 2, 36). Mortality data suggest excess deaths amounting to several hundred thousands, primarily in Galicia (present Poland) with excess deaths of up to 300,000 and in Bohemia-Moravia (present Czech Republic) with excess deaths of up to 100,000. Hunger typhus was a major cause of death. The Wadowiczer Comitat in Galicia (then under Austrian rule, present S Poland), counted 60,000 to 80,000 deaths in 1847 due to hunger and typhus, the typhus epidemic hitting somewhat earlier than in neighbouring Upper Silesia (Virchow, 1848: 38). In Austrian Silesia thousands 'had nothing to eat except grass and nettles, coltsfoot, clover and blood'. In the big cities of Austria maize and/or clover were added to the wheat flour (Macartney, 1968: 313).

Belgium counted nearly 4,000,000 inhabitants in 1846/50. From 1845 to 1846 a famine raged and many people died[249]. Death rates exceeded birth rates in several communities. The province of West Flanders recorded a peak death rate of 3.93 per cent in 1847 (Hofstee, 1978: 199). In Belgium the death rate was highest in 1847 with 2.77 per cent against an average of 2.33 per cent over the period 1841/5, suggesting excess deaths of ⩾35,000 from hunger and disease (André & Pereira-Roque, 1974). All of Flanders was said to have ~50,000 deaths in excess (Ó Gráda, 1989) including the ~12,000 killed by typhus in 1846/8 but excluding the nearly 6,000 deaths by the cholera epidemic of 1848 (Lamberty, 1949: 136).

France. Marriage and birth rates decreased and death rates increased. Population growth slowed down considerably. Migration from the countryside to the cities intensified. Emigration exceeded immigration (Table 3.2).

The Germanies. Excess deaths[250] in the Kingdom of Prussia can be tentatively estimated at ~140,000. The province of East Prussia recorded 40,000 deaths from disease, winter 1847/8 and spring 1848. In Upper Silesia (Prussia, present Poland) ~80,000 people suffered from hunger typhus, with ~16,000 deaths (others say 30,000), and ~50,000 deaths in total. Data about numbers and diseases are not quite consistent.

Though this not the place to discuss the medical aspects of the hunger epidemics that followed the harvest failures, one exception has to be made. Early in 1848 the Prussian Government had sent R. Virchow, a young and ambitious surgeon at the Charité Hospital in Berlin, to Upper Silesia to investigate the ongoing typhus epidemic. He wrote an extensive scientific report (Virchow, 1848) in which he left little doubt as to the nature of the major disease, which we now call typhus (caused by a *Rickettsia*). Virchow was as critical of the Prussian government as the circumstances allowed. He referred to an anonymous brochure on 'The hunger pest in Upper Silesia' (Anonymous, 1848a), which really is an indictment of government policy. Besides a high scientific motivation Virchow had a strong social interest.

Switzerland. During the blight years the marriage and birth rates went down, but the death rate was not affected (Bickel, 1947: 123, 148).

3.3.3. Pauperism

Pauperism stands for the chronic poverty of the lower classes in a region and/or period[251]. Toward the mid 19th century pauperism was widespread in Europe, where many states slid into economic recessions in the 1840s. Rural pauperism was a rather new phenomenon with causes differing according to region[252]. (1) Wages in rural areas hardly increased during the first half of the 19th century, probably because of the relatively rapid population growth, as in the Netherlands, Belgium (Flanders) and Prussia. (2) The rise of modern textiles industry caused a sharp decline of the cottage industry, as in Austria, Flanders and Prussia. (3) Division of the commons increased the gap between estate owners and peasant farmers, as in Prussia. (4) Extreme parcellation of land led to great poverty among peasant farmers as in SW, M and W Germany, and in Switzerland.

The bad times caused a decrease in demand and so the craftsmen, then quite numerous in the villages and rural cities, went out of business. They too had to apply for relief or to go begging. Continent-wide the behaviour of the poor followed more or less the same pattern (Box 3.3).

The Netherlands. In a period when class distinctions were an accepted phenomenon pauperism in itself was not rejected. Only revolutionaries could foster different ideas. The 'haves' should be nice to the 'have-nots', do relief work, and feel good. In 1844 the country experienced a slump, factories were closed. Labourers lost their jobs and became dependent on poor relief funds (Brugmans, 1929: 216). In the year 1846 the rust, voles, drought and blight boosted food prices and so deepened poverty, increased unemployment, and awakened slumbering unrest in the cities.

Begging[253] and theft increased considerably from 1845 to 1848. Hungry people roamed the streets. Those having something to loose became afraid of group begging, with extortion under threat. The police had much more work to do. In some parts of Zeeland more convictions for begging were pronounced in 1846 and 1847 than in all twenty years before[254]. In the relatively prosperous province of Friesland over 14 per cent of the rural population was on the dole and, in the area with arable land mainly, over 20 per cent (Van Zanden, 1985).

Austria. Industrialisation led to underemployment and impoverishment on the countryside, with pauperism of a 'rural proletariat' (Rumpler, 1997: 255) as the result. Many migrated to the cities where, eventually, their situation became worse than ever.

Belgium. Over 400,000 Flemish persons lived off the textile industry (Lamberty & Lissens, 1951). On the Flemish countryside the cottage industry, mainly linen spinning and weaving, was yielding to industrial production, ensuing loss of income and purchasing power. In one Flemish parish with 225 households a hundred had to go begging (Lindemans, 1952: 189). Bands of people marauded the fields and roamed the streets, begging and stealing. Windows were smashed with the sole purpose to be incarcerated and fed (Lamberty, 1949: 136). The Walloon provinces in Belgium were hardly affected because of their industrial expansion.

France. 'Endemic misery' reigned during the 1840s[255]. Land owners had begun to invest in industry and movables disinvesting from agriculture[256] (Droz, 1967: 267). The French estate owners began to loose their grip on the peasantry[256a]. Peasants and landless poor obstinately struggled for independence by all means, legal and illegal, as vividly depicted by the novelist

---

Box 3.3. The fate of the poor during a famine, as stated by various authors.

The patterns described here were common to most *crises de subsistance*. The crisis sparked by the potato murrain was the last in Continental Europe (Scandinavia and the Soviet Union excepted).
The flour for bread making, if available, was mixed with several admixtures among which bran, clover meal, *Faba* bean meal, and meal[257] of queck grass roots (*Elytrigia repens*), maybe also with meal from wood or hay as in other famines.
Most decent people went through a period of denial, sparing whatever food was available. The size of the meals was reduced, and then the number of meals was reduced to two or one per day. Whatever money available was spent on food and house rent and no goods were bought. Jewelry, cloths, furniture, household utensils and professional equipment were sold. Families moved to cheaper dwellings. Desperate people pilfered bakeries and food stalls. Members of town families roamed the countryside to buy food at usurious prices.
Country people, who anyhow had little to sell, tightened their belts. If they were out of regular food they collected whatever eatables they could find, browsing the fields, stealing from standing crops, searching the refuse piles and dunghills, stealing sheep or goat, killing dogs, cats and rats, slaying diseased animals, and digging for buried animals. Grass, nettles, clover, coltsfoot (*Tussilago farfara*), fungi, and the bark of trees were on the menu. Some items, including rotten potatoes, were eaten raw but, preferably, they were cooked in a stew that got thinner and thinner.
The able-bodied tried to find work, often relief work, e.g. heavy digging at low wages. The disabled, by hunger and sickness, died rapidly from hunger typhus or slowly from intestinal afflictions, tuberculosis, or physical exhaustion. Farmers dropped dead in their fields[258].
Many jobless people found food with relief organisations, parishes, municipalities, private institutions, often created for the purpose, or with private persons. Relief funds were always too limited, and they sustained people partially and for a restricted time only, until the funds were exhausted. 'Relief fatigue' with the sponsors was mentioned several times.
Setting aside their pride, people went begging, first individually, sometimes following a prescribed route through town. Bands of beggars were formed by men, women, even children. These bands were threatening and meant to threat. They visited farmsteads and, if they did not get their alms, they might set farm or mill afire. They stopped the grain traffic on the roads intercepting and plundering the grain-loaded carts.
Crime rate soared because of the many petty thefts. The 'haves' on the countryside became afraid and fear spread over the country. The police, sometimes assisted by the military, cold no longer meet the situation. Vagabonds were happy to be arrested, because in jail they were fed at least. Once in jail, or in the workhouse, they ran the risk of dying from typhus or dysentery, or being infected by tuberculosis.
In the typhus-stricken areas thousands of children were orphaned. If lucky, they were found, placed in an orphanage, washed, given clean cloths, fed and taught. They had a fair chance to survive and start a new life, without relatives.
On the countryside violence was frequent in France, quiet resignation dominated in the Netherlands and in Belgium, inertia in Silesia, and sullenness in Austria.

Honoré de Balzac[259] in '*Les paysans*' ('The peasants'). A departmental governer (F: *préfet*) in the SW reporting on the winter 1845/6 said that at one town 'Over 25,000 come to the point of having nothing left to eat' and at another town two thirds of the inhabitants 'lack every thing, no money, no bread, no potatoes and, finally, no credit'[260]. The catastrophic grain and potato harvests of 1846 caused widespread distress (except, maybe, in S France) and great social disturbance[261].

The Germanies gave a similar picture. In the neat archduchy of Baden people in town and countryside lived soberly, in mutual dependence, without many reserves. The failure of the potato and grain harvest in 1846 caused a famine and an economic crisis among all strata of society (Real, 1983). No wonder that 'decent artisans' without work or money went begging, to the amazement (and irritation) of the Dutch reverend O.G. Heldring (1847) vacationing along the river Rhine in 1847, at cherry picking time, just before the grain harvest.

3.3.4. Hunger riots

Throughout history the common people, when threatened in their subsistence by food scarcity, had only one way to show their discontent to the ruling authorities, rioting. Food riots are of all times.

The Netherlands. Generally speaking the Dutch rural workers were a subdue lot (Terlouw, 1971: 290), suffering in silence. 'Labourers were powerless and slow, although not unwilling' (Brugmans, 1929: 169). They had neither voice nor representation. On Sundays they were told to trust in God. Pilfering was rare and severely punished. Nonetheless, quiet, complacent Holland had its 'potato riots' in September, 1845, when townspeople protested against the high potato prices. An eye witness mentioned the 1845 riots in Delft, where groceries and bakeries were looted (Van der Hardt Aberson, 1893). Riots also occurred in Leiden and even in The Hague, the seat of the Dutch government.

At the time, the Netherlands had changed from a grain eating to a potato eating nation, with an annual production in the early 1840s of ~1 million tonnes (~14 million hectolitres). In the first blight year, 1845, the national production was less than 0.3 million tons (<4 million hl), a loss of over 70 per cent. Grain was rapidly imported and the consumption of legumes increased. The winter of 1845-1846 was mild, funds for the relief of the poor had plenty of money, and relief-work was organised. Things were under control.

Unfortunately, the blight struck again in 1846. The severe and long winter of 1846/7 added immensely to the misery of the cottagers and jobless townspeople. Unrest flared up again in June, 1847[262]. In the port of Harlingen a ship with destination England, the 'Magnet', had been loaded with potatoes (Jansma & Schroor, 1987: 59), possibly from the early varieties of the 1847 harvest. A mob protested, molested officials, and turned to looting. The insurrection spread to the towns of Leeuwarden and Groningen. In Groningen the dragoons had to restore order at the price of five casualties (Van der Heiden, 2001). Radical politicians tried to exploit the general discontent, with little success.

'Liberals' came into power in 1848. Concerned about the social unrest, they arranged investigations into the morality of the working class. Provincial reports were published in 1851 (Van Zanden, 1991). The results were gloomy as e.g. whole families taught their children to beg and steal firewood (as in Zeeland, *ibid.*: 12). The liberal answers were not oppression and charity but good upbringing and school education, in the spirit of the Enlightenment.

Austria. Unrest occurred everywhere in the Austrian realm, in the major cities but also on the Polish countryside, where national politics and nascent liberalism added to the general dissatisfaction with Austrian rule, enhanced by the dearth of the times. In the suburbs of Vienna bakeries and market stalls were plundered (Lutz, 1985: 944).

Belgium. The rural population of Flanders appeared to be an 'amorphous mass'. As an exception, a few bakeries in the city of Ghent were pilfered (Lamberty, 1949: 136).

France. In France the poor harvests of 1846 caused a slump in consumer demand that triggered an economic crisis. Gangs of hungry people roamed the countryside. Farmers, millers and grain merchants were threatened with arson[263] and extorted, sometimes by masked persons. Grain transports were hold up, often by enraged women[264]. At times the rural masses burst out in fury (Agulhon *et al.*, 1976: 78), with destructions and smashing up of tax archives. Smouldering unrest finally exploded in the February revolution, Paris, 1848 (see below). France was the only country where rural unrest contributed to radical political change.

The Germanies. In the German speaking countries hunger riots (G: *Hungerkrawallen*) occurred in the spring of 1847 in Berlin[265], Hamburg, Stuttgart, Ulm, and various other towns (Lutz, 1985: 944; Wehler, 1987: 657).

3.3.5. Migration

Nothing in Continental Europe paralleled the massive emigration from Ireland to North America (~1,000,000) and England (~300,000) during and after the hunger years. Nonetheless, emigration to the Americas was frequent for economic, religious or political reasons. Disillusioned and police-threatened political activists sometimes went digging in the recently found (1848) goldfields of California.

Many in Continental Europe packed up and moved, or simply went adrift. Migration from the land to the city is of all times, but it was intensified under the pressure of the circumstances. People in search of work, that was hardly available in the 1840s, added to a new urban proletariat, jumping out of the frying pan into the fire. Transmigration within and between W European countries was frequent, as e.g. from Flanders (Belgium) to the industrial area of NE France.

The Netherlands. Discontented families emigrated to North America. Poverty, lack of prospects, and religious dissidence were among the reasons. Emigration peaked in 1846 and 1847, but we speak of only thousands of Dutch[266], not the hundreds of thousands of Irish. Migration from countryside to towns may have intensified in the late 1840s.

Austria. Impoverished country dwellers flocked into the cities in search of work. In Vienna they only met with more misery and awful housing conditions.

Belgium. In Belgium, an exodus took place from hungry Flanders to the booming industrial areas of francophone Belgium and N France. The numbers involved are estimated at tens of thousands (Lamberty, 1949: 136).

France. France experienced an intensified 'rural exodus' of day labourers, craftsmen, construction workers, and outworkers to the cities amounting to ~800,000 persons over the period 1831/51

(Agulhon *et al.*, 1976 p80). Net emigration over the years 1846/51 was ~276,000 (Armengaud, 1971). Many emigrants went to Mexico, North America, and Algeria. Thousands of Germans, stranded penniless at the port of Dunkirk, were shipped to Algeria by the French authorities.

The Germanies. Emigration to North America in the years 1840/4 averaged ~22,000, in the years 1845/9 ~62,000, increasing in the 1850s to hundred thousands (Marschalck, 1973).

Switzerland. There was immigration and emigration but the numbers were feeble.

3.3.6. Official relief

The new liberal thinking with its *laissez-faire* was not much in favour of relief, at least at state level. Governmental actions were usually limited to adjustment of levies and taxes with a view to limit the price of rye (the food of the poor), by facilitating grain imports and, sometimes, reducing exports. Intermediate levels (province, department) showed more readiness to act. At municipal level many authorities, in face of the want of their citizens, felt compelled to take action. Churches had relief systems stand-by. Private persons, individual or organised, readily stepped in to help (§3.3.7).

The Netherlands. The Dutch Parliament discussed the withdrawal of the Dutch Corn Law dispassionately, technically[267]. Import duties on food commodities were suspended, first temporarily (§3.2.4), later definitively. The provincial governors were not allowed to provide money for relief, not even on loan. The municipalities were discouraged to provide relief but, inevitably confronted with the misery, they used existing poor-relief funds. They provided public works, food, and price control of rye bread (the affluent ate wheaten bread). They reduced municipal bread excises and milling taxes, and even went so far as to subsidise rye bread. Soon their funds were exhausted. Some municipalities tried to supplement their funds by imposing a levy on the well-to-do. Generally speaking, the municipalities did what they could do. In the province Noord-Holland some 25 per cent of the population lived on the dole in 1848 (Kossman & Krul, 1977: 71).

The Dutch Government showed its compassion by declaring 2 May 1847 a national day of prayer (Van der Heiden, 2001). There is no evidence of any effect other than keeping the poor quiet and poor. The day yielded a collection of sermons that showed how wide a social gap stood between the reverends, 'haves' usually, and their suffering flocks[268].

Austria. Some measures were taken to avoid a market catastrophe and a famine but public works had to be stopped because the government went out of money (Macartney, 1968: 313). The poor were referred to the parishes and to private charity. In Hungary hundreds of lives could have been saved had there been adequate roads to transport grain to areas in need (Elsner, 1847: 811).

Belgium. The state provided credit for work, purchase and transportation of grain, and subsidies on bread prices (Abel, 1974: 379). Communal bakeries, bread coupons, and public kitchens were among the means to reduce the misery in the towns but they were of no avail on the countryside (Pirenne, 1932: 129). In 1847 about one third of the Flemish people lived on charity (Lamberty, 1949: 136).

France. Departments and municipalities sometimes voted money for public works and support of the poor[269]. Cities such as Toulouse and Bordeaux distributed bread coupons. At the local level, the poor often received municipal or private support.

The Germanies. Münter (1846) gave an overview of immediate actions taken in 1845, among them public works, credit, export bans, and the opening of military stores. In Upper Silesia official relief was late. In the Rybnik area one third of the population received aid in the form of a pound of meal per person per day (Virchow, 1848: 27). In the Rhine Province the city of Cologne, not a poverty-stricken town, had 25,000 out of the 95,000 inhabitants registered on the poor list (Lutz, 1985: 116).

Ireland. Robert Peele, Prime Minister of England during the early phase of the Irish Famine, was a conservative with rather liberal ideas. Facing the emergency he did what he could do[270]. He bought maize in the American colonies to provide for the hungry. His successor (June, 1846), lord John Russell, stopped grain imports and public works, but in 1847 he felt compelled to import food for the 'soup kitchens'.

Poland. The distress was great, which is no surprise in view of the difficult transportation over large distances (Elsner, 1847: 810).

3.3.7. Private relief

Private relief was very active and took many shapes. Existing organisations, primarily the churches, usually acted first. Citizens organised themselves in *ad hoc* societies, or took action individually. Here follows a somewhat arbitrary selection of examples.

The Netherlands. In the Netherlands charitable institutions, primarily the churches, were very active. In the course of 1846, unfortunately, their funds became exhausted. During the mild winter of 1845/6 private persons, farmers and manufacturers, could provide work so that the poor labourer could earn some money. Most of this work was for improvement of roads, waterways, and ditches. The severe frost of the winter 1846/7 made such work impossible and the poor were left without income (Hooijer, 1847; Terlouw, 1971: 288).

Of course, a national campaign was organised for the relief the poor, though too late to prevent early deaths (Bergman, 1967: 400). Private initiatives sprang up unexpectedly. Towards the end of 1845 some wealthy retired tradesmen in Amsterdam bought a few shiploads of grain and sold the grain at reasonable price in an attempt to lower grain market prices[271], the opposite of hoarding. In the small Frisian village of Irnsum '... the bread for the common folks had been down-priced to five pennies, for which the expenses are found in a collection among the most dignitary residents'[272]. In Dokkum, a Frisian town, a committee collected 'generous gifts' to provide the poor with reduction coupons so that they could buy their rye bread at a fixed price below the current retail price[273].

Austria. The poor had to rely on parishes and on private charity. Some estate owners made food from their stores available to the villagers, other did not.

France. In many places private money was collected in order to avoid food shortages (Houssel, 1976: 234) and to keep the people quiet. In the town of Lille a private association distributed aid in cooperation with the town's welfare office (Jardin & Tudesq, 1973: 236).

The Germanies. In the severely affected area of Upper Silesia, where the officials did not want to see the misery, the 'Breslau Committee', which first had to beg for money in all of Germany, was on the spot well before the government!, wrote a bitter Virchow[274]. The very active chairman, Prince Biron von Curland, contracted typhus and died. In the city of Koblenz, 1849, a 'Society for the Procurement of Cheaper Food' bought Russian grain in the ports of Amsterdam and Rotterdam.

Ireland. Shear numbers made relief a nearly hopeless enterprise but various charities took action. The 'Society of Friends', the Quakers, operating from England, should be mentioned explicitly (Bourke, 1993: 177, 182; Woodham-Smith, 1962).

3.3.8. Access to credit

In today's developing countries access to credit for the poor, in the form of micro-credit, is a hot item. Around the mid-19th century Continental Europe was a developing world where (micro-)credit would have been useful, as proposed by some enlightened thinkers. Early initiatives in England led to the establishment of the first co-operatives.

The Netherlands of 1845 hardly participated in these developments. Social motivation, so successful in the 20th century, was scarce among the 'haves' of the mid 19th century. O.G. Heldring (1845: 21), parson in a hard-hit area, discussed the need of small local credit banks for the poor. His plea remained without response. Ph.A. Bachiene, a tax administrator in Sluis (Zeeland), worked hard to organise small loans (D: *kleine voorschotten*) for the poor with remarkably little success (Bouman, 1946: 136). Micro-credit was not a successful issue (De Bosch Kemper, 1851: 258). At the time, the Netherlands were not yet ready for modern banking, borrowing money here and lending it out at a premium there (Brugmans, 1929: 71).

Austria. Credit was one of the themes discussed at the Tenth Meeting of German Farmers and Foresters in Grätz (Styria), 18 September 1846 (Mentzel, 1848: 108).

France. The head of the 1848 interim government, General E.L. Cavaignac, wanted to provide cheap credit to peasants. His proposals were rejected by the Assemblé Nationale (Newman & Simpson, 1987: 897). Cavaignac had some modern ideas. He wanted to promote producer co-operatives and farming schools.

The Germanies. In the province Prussia the estate owners could modernise their agriculture because they had access to credit. A warm plea for credit to small farmers was made by Lette (1848), after a visit to the province Prussia in 1846[275]. But it was F.W. Raiffeisen, mayor of a rural town in Westfalia, who initiated farmer co-operatives, inspired by the misery of the late 1840s. He wrote[276] 'As to the history of the loan societies, their birthplace is the lower Westerwald, in the Prussian Rhine Province, the time proper of origin the emergency year 1847'. What began as an informal 'consumer society' (rather a charity committee), that provided cheap bread and later seed potatoes for the 1847 planting season, developed gradually into rural banks, strictly local, non-profit, governed by village notables. Later these village-level banks were knitted together into the German Raiffeisenbank. Neighbouring countries followed the example. In the Netherlands the RABO-bank (the RA remembers Raiffeisen) has become one of the largest Dutch banks, still basically a co-operative, a bank with AAA status.

3.4. The economic depression of the 1840s

3.4.1. Mercantilist attitudes

The mercantilist policy of the 18th century, implying protection of the national production against cheap imports by means of tariff walls, was continued by several nations until far into the 19th century. One example were the English 'Corn Laws' protecting the English grain producers by means of flexible import duties, high when grain prices at home were low and low when they were high.

The Netherlands followed the example by the law of 29 December, 1835, 'to promote the interests of agriculture', introduced after fierce resistance of the grain trade. A relaxation of this protectionist measure occurred by the law of December 18th, 1845, under the pressure of the grain merchants, and in recognition of the shortages caused by the potato blight (Terlouw, 1971: 276/7).

Austria. Austria distanced itself from the German Federation and maintained its tariff walls (Wehler, 1987: 131).

Belgium, independent since 1839, had a commercial regime with few restrictions to international trade (Pirenne, 1932).

France. The Emperor Napoleon III mitigated the mercantilist regulations only around 1860.

The Germanies. The German Federation (G: *Deutsche Bund*) had a comparatively liberal Customs Agreement (G: *Zollverein*), promoting within-federation trade and facilitating international trade (Wehler, 1987: 126/7, 557). Prussia was a wheat exporter.

3.4.2. The 'commercial crisis' in NW Europe, 1847-1848

An eye-witness of the 'commercial crisis' sketched the 'career of the crisis' as a three tier process (Morier-Evans, 1848). (1) The 1840s were the hey-days of railway construction. A 'railway-mania' reached its peak in 1845 but in October of that year a panic in the share-market punctured the bubble leaving many speculators penniless. (2) The poor harvest of 1846, of potatoes in particular, necessitated the purchase of food abroad. Money was tight, interest rates became high. Grain speculation was rampant[277]. Corn prices declined in May, 1847, and caused the 'food and money panic'. (3) On top of all this came the 'French Revolution' of 1848 causing a loss of trust. Asset values plummeted on the Stock Exchanges of London and of the Continent[278]. The list of failing traders, merchants, money dealers and bankers is long. Crisis sprawled over the Continent. Investment in railways, an investment in new communication comparable to the ICT hype[279] of the 1990s, became risky business in continental Europe[280].

The English treasury remained in fair shape but several continental treasuries were in a critical state (Lutz, 1985: 234/5), due to warfare and/or poor management, as in the Netherlands,

Austria, France and Prussia. The continental states had great difficulties to meet the financial crisis. The Netherlands and Prussia, floating big loans, made a narrow escape.

The industrialisation of England in the first part of the 19th century was not matched on the Continent. Rapid population growth and the dissolution of the cottage industry produced a supply of labour that could not be absorbed by the national economies of the Continent. Pauperism (§3.3.3) and the rise of a rural as well as an urban proletariat were among the consequences.

On the Continent the harvest failures caused a steep rise in food prices, especially in the towns, and this entailed a sharp dip in the demand for industrial products that, in turn, led to dismissal of personnel, more unemployment and more poverty. The young surgeon Rudolf Virchow in Berlin wrote to his father[281] 'every day new masses of manual workers loose their means of support, the factories close down one after the other and all of us are disturbed in our livelihood'.

The Netherlands. At the time the Netherlands were a somewhat self-contained nation, relying on agriculture and commerce, but with little industry. The country was on the verge of bankruptcy (Bergman, 1967: 417). In the fall of 1845 speculation in grain caused a temporary lack of capital (Terlouw, 1971: 290). Factories came to a stop and unemployment rose rapidly. Between January and April, 1848, Dutch shares lost ~40 per cent of their value, several foreign shares over 50 per cent (Anonymous, 1849: 216). At least a dozen Amsterdam money dealers of good renown had to suspend their payments (Morier-Evans, 1848: 119).

Austria, spending much money on the military, endured an acute financial crisis in 1847/8. Grain speculation and hoarding boosted food prices, and because of the high food prices people could no longer pay their taxes and other debts. Due to the lack of purchasing power of the public several manufacturers could no longer sell their products, dismissed their labourers and shut down. The famous bank Rothschild at Vienna, the Emperor's financier, stopped its payments on 6 March, 1848, thus informally acknowledging the bankruptcy of the Imperial government (Lutz, 1985: 99; Rumpler, 1997: 276). A run on the banks occurred. Share values on the Vienna stock market plummeted, especially of the railway shares.

Belgium. The Paris revolution caused a rush on the banks; share values were halved (Pirenne, 1932: 131). The government, however, stayed level-headed.

France. Again, a three-tier process was discernible with (1) a crisis in construction work and railways, followed by (2) a crisis of confidence, deepened by (3) the agricultural crisis (Jardin & Tudesq, 1973: 237/8). The catastrophic potato and grain harvests of 1846 increased the food prices, forced up by speculation, and caused a sudden fall in the demand for industrial products, leading to unemployment[282]. The nation slid into an acute depression, thought to be due first and foremost by the harvest failures (Furet, 1988: 374). The 1847 harvest was relatively good but made no end to under-consumption. Bankruptcies and unemployment spread, shares dropped in value, credit became expensive or unavailable, and trust was lost[283]. Some banks had to stop their operations (Braudel & Labrousse, 1976: 375).

The Germanies. Prussia experienced an economic crisis due to population growth, massive unemployment, and extensive pauperism[284]. The failed potato harvest of 1846 deprived the poor from their staple food, not to mention the poor rye harvest of that year (Chapter 2). Food and credit became expensive, unemployment increased, demand fell, industries and crafts suffered. The Frankfurt Stock Exchange plummeted overnight. A highly respectable bank, the *Schaafhausensche Bankverein* at Cologne, went bankrupt and closed its doors in March, 1848, dragging along some 40,000 clients. The archduchy of Baden reached a state of emergency (Riegger, 1998).

3.4.3. Economic liberalisation

Clearly, the authorities in many countries could no longer handle the explosive combination of circumstances. Many of those in power adhered to classical liberalism with its *'laissez faire'* as the leading doctrine. In their hands it was a cruel doctrine since the common people were left to starve whereas some persons made much money in hoarding, grain speculation, wheat exports (e.g. Prussia), and grain supply to distilleries (Prussia again). Brandy was the last consolation of many poor and hungry people. Rudolf Virchow having visited Upper Silesia in early 1848 wrote[285] 'the child at its mother's breast is already fed with brandy'.

The Netherlands. The Finance Minister F.A. Van Hall, who had just saved the nation from bankruptcy by floating a large loan[286] at modest interest, is sketched as a sturdy non-interventionist. He kept his stand against the pressure of many, even of King William II. In Parliament, however, he defended the bill to suspend import duties of agricultural commodities in order to stimulate the importation of potatoes and grain[287]. As the high food prices asked for a concession by the Dutch Government, the Dutch 'Corn Laws' were suspended by law of 18 December 1845 (§3.2.4). The suspension was prolonged until 1 October 1847 since the bad harvests of potatoes and rye in 1846 accentuated the importance of a free grain trade (Terlouw, 1971: 283). Thus the turn from protectionism to free trade, a first victory of economic liberalism in the Netherlands, was made thanks to blight and rust. There was no way back (Terlouw, 1971: 248). In a similar vein Robert Peel, the conservative Prime Minister of England, was induced to abolish the English 'Corn Laws' in 1846, as the dearth of food threatened the livelihood of the industry workers.

3.4.4. Causes of famine

An extensive literature exists on the famine as a socio-economic phenomenon[288]. Let it suffice here to recall that two major causes have been identified. The first is a strong 'food availability decline' (FAD), obvious in the years 1845 and 1846 with their harvest failures. The second is the lack of purchasing power to buy whatever food is available. Simplified, the farmers had nothing to sell, they had no money to buy services of craftsmen or to purchase industrial products, so that the manufacturers could not sell their produce and had to close down. The jobless craftsmen, construction workers and industrial labourers could no longer pay for food that rapidly increased in price (Figure 3.8). In the late 1840s the two causes were clearly interrelated.

The actual situation was far more complicated and should be considered in terms of countryside versus town, land owners versus day labourers, farming community versus craftsmen in villages and rural towns, and so on. In some areas, among which E Prussia, food exports and hunger existed nearly side by side.

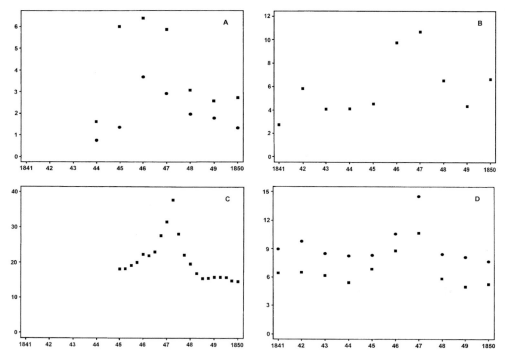

Figure 3.8. Prices of major food commodities on the European Continent to illustrate the dearth in the years 1845-1847.
(A) Prices of late potatoes (maximum and minimum) at Leiden, Netherlands, in Dutch Guilders per mud = Dutch bushel = ~70 kg (after Terlouw, 1971: 307). (B) Potato prices in Zürich, Switzerland, in Swiss Francs per 100 kg (after Brugger, 1956: 114). (C) Wheat prices in France, in French Francs per hl (modified after Abel, 1974: Fig. 69). (D) Wheat (●) and rye (■) prices in the Netherlands in Dutch guilders per hl (Staring, 1860: 586).

3.5. The events of 1848

3.5.1. Political aspects

The Treaty of Vienna (1815) reorganised Continental Europe after the Napoleonic Wars, restoring monarchism in old and new nations. Rulers of several reconstructed German principalities donated constitutions to befriend their, often new, subjects[289]. The Netherlands were a newly created nation composed of the present states of Belgium, Luxemburg, and the Netherlands. In 1815 it became a constitutional kingdom with an agrarian north (present Netherlands) and an industrialising south (present Belgium and Luxemburg). In 1839 this nation fell apart. Belgium proclaimed independence and found a king willing to rule under its new and very liberal constitution of 1831. Luxemburg became independent in 1867.

In many capitals intellectual life was fermenting in the 19th century. The Enlightenment of the 18th century had its political culmination point in the Great French Revolution of 1789. Two main streams of political thinking built on that revolution's credo 'freedom, equality, fraternity'. One stream, leading to socialism and communism, sought to realise its ideals by propagating common property of production means. Many of its proponents addressed the street, wanting revolution. Marx & Engels published their 'Communist Manifesto' in 1848! Another stream, liberalism[290], wanted civil rights for individual citizens under protection of the law. Supported by entrepreneurs, middle class people, academics and, often, high-placed civil servants, it sought to attain its objectives by evolution. A third stream of political thought, nationalism, was fed by Romanticism. These interconnected streams formed an explosive mixture.

The Netherlands. In the Netherlands the government suppressed all unrest, primarily the relatively innocuous hunger riots, with undue force. When a crowd had collected on the Dam Square in Amsterdam, in front of the Royal Palace[291], King William II felt the threat of revolution and apparently panicked (Jansma & Schroor, 1961: 61). A conservative, he became a liberal in one night. He invited a Member of Parliament, Professor J.R. Thorbecke, to chair a committee and write a new constitution. This constitution, for the time quite liberal, was written within four weeks (Van Schie, 2006). Though liberals were a small minority in Parliament, the Constitution was accepted in 1848, not without pressure by the King. The Constitution was so modern that, with some inevitable modifications, it remained basically unchanged until today[292].

Austria. Protests and demonstrations in Vienna, capital of the Austrian/Hungarian double monarchy, followed the hunger winter of 1847/8 and, in the end, induced the Emperor to proclaim a Constitution in 1848.

Belgium. The liberal constitution of 1831 stood out as an example. Thanks to its liberal climate several revolutionaries flocked together in Brussels, among them Marx and Engels, during the late 1840s. In 1848 an incipient revolution was crushed by the Brussels police (Eskens, 2000: 102/3). A relation between the revolutionary spirit in Brussels and the rural misery, in Flanders especially, was not found.

France. The general discontent of the 1840s burst out in violence, first in Paris, on 22 February 1848. The Bourbon King was chased away on 24 February. On 10 December 1848 Prince Louis Napoleon Bonaparte was chosen President of the Second Republic[293]. 'The collapse of authority unleashed a massive wave of direct action in towns and villages where capitalist industry and agriculture and state politics had threatened livelihood' (Newman & Simpson, 1978: 897). The causal relationship between harvest failures and revolution is most obvious in France (§3.3.3).

The Germanies. General discontent 'had shaken the authority and credibility of the State'[294]. The spark of revolution jumped over to Berlin (Prussia) where it ignited the ´March Revolution´, which was suppressed by force. Nonetheless, Prussia received its first constitution in 1848. Events were milder in the kingdom of Bavaria and the Archduchy of Baden (Riegger, 1998), in the same year.

Hungary. Hungarian liberals, revolting in Budapest, wrote and accepted a liberal constitution for Hungary in 1848.

Ireland. Inspired by the news from the Continent a small band attacked the police station of Tipperary but this minor rebellion was crushed by the police[295]. Taking a long shot, Ordish (1976: 138) commented 'The blight ... nourished the seeds of the Republic of Eire'.

Other countries. Revolts with a more nationalistic touch occurred in several city-states of present Italy, 1848, Prague (Bohemia, then under Austrian rule), Poznań (in present Poland, then under Prussian rule), and several other places, without much success. Switzerland, a confederation of relatively independent democratically governed cantons, made a special case. After serious inter-cantonal quarrels a liberal constitution was accepted in 1848, notwithstanding the admonitions of the pious mister Gotthelf (1847)[296]. King Fredrick VII of Denmark announced the introduction of parliamentary democracy on 20 January 1848, leading to the constitution of 5 June 1849. England, with its archetype of a constitutional monarchy, Spain and Russia remained untouched by the 1848 revolutions.

Counter-revolutions. 'The revolution of 1848 was not the product of an intentional revolutionary action: it was rather the implosion of a traditional form of government on the European continent, of which the legitimate validity had collapsed and of which the fight for its maintenance had at once been felt as hopeless' wrote Mommsen (1998)[297]. Many new rights had been obtained by the people in 1848.

When the worst was over countervailing forces tried to undo the results of revolution by means of 'counter-revolutions'. In Prussia and Austria and their dependent areas, and in various German principalities, rulers cancelled several of the newly accorded rights, 1848 or 1849. France followed in 1851. Some innovations, however, survived such as voting rights for at least part of the male population, and other civil rights. Serfdom was abolished definitively. Economic power was transferred from the nobility to the middle classes, an irrivocable change 'from feudalism to capitalism' (Gieysztor *et al.*, 1979: 417).

3.5.2. The potato blight and the revolutions

An epidemic of potato late blight was followed by an epidemic of constitutions, more or less liberal. Was there a causal relationship? Yes and no!

No, since the roots of liberalism reach far beyond 1844 when the blight struck first. No, as the general discontent in the capitals, towns and rural areas had built up in the 1840s due to a variety of reasons, economical and political.

Yes, as the fall of share values in 1846 was not only coincidence; it was also due to the high price of credit needed to finance large grain imports. Yes, because the dearth and famine following the poor potato and grain harvests deepened the discontent (Riegger, 1998). Hunger makes rebels and insurgents. Yes, because the blight and, again, the poor grain harvest in France, where the revolts began, triggered an acute economic depression that led to revolutionary unrest.

I quote a few authors with different backgrounds:
- Wehler (1987), a German historian, had some reservations[298]. 'To assume a direct connection between hunger crisis, grain riots and potato revolts on the one hand, and the revolution of 1848 on the other hand, would be to prefer a simplifying shortcut to more complicated interrelationships' but he considered 'the crisis from 1845 to 1848 the prelude of the 1848 revolution in all of Europe'.
- Pirenne (1932), a Belgian historian, stated[299] 'the economic crisis was soon complicated by a food crisis'.

- The German historian Lutz (1985: 244) wrote[300] 'The structural economic crisis of the forties, characterised by population pressure, massive unemployment and pauperism, was sharpened by an acute crisis in 1846/7. Potato blight and harvest failures led to a famine manifested by epidemics and a reduced birth rate', (*ibid.*:245) 'So grew, from famine and hunger epidemic, oppositional criticism and public awareness formation' and (*ibid.*: 245) 'A direct relationship between the hunger unrest 1847 and the revolution was also assumed by non-socialist contemporaries'.
- The Dutch publicist Van der Heiden (2001) stated 'The story of the, after all peacefully achieved 1848 revision of the Constitution ... should, strictly speaking, begin with *Phytophthora infestans*'. Apparently he followed the Dutch historian of agriculture Sneller who wrote 'it [the potato blight] gave an additional impuls to the peaceful revolution of the national government in the year 1848'[301].
- And the French historian Le Roy Ladurie[302] said in one of his inimitably intricate but highly precise sentences 'Raged after all, always with accidentality and variability, the long and hot drought of the spring and summer of 1846, which, with the potato disease, notably in Ireland, carries certain responsibilities by way of the economic depression of 1847 born from that poor harvest; she implies effectively a kind of climatic guilt, though partial, in view of the ultimate outbreak of the revolutions in France and then in West and Central Europe, beginning February, 1848, in the environment of an economic crisis that indeed sprouts from the difficult post-harvest year 1846-1847; in the environment, also, correlating, of a certain discontent of the various populations'.

The number of excess deaths due to potato blight, failed grain harvests, and corollary diseases, 'hunger typhus' foremost, cannot (yet) be established with any accuracy. Continental Europe suffered a terrific blow with ~700,000 victims (Table 3.3), a disastrous number approaching the Irish catastrophy, though spread out over a larger area with a larger population.

3.5.3. The fate of the farmers

Though the fate of the farmers in the 'Black Years' is not my topic a few words may be said. Landless labourers and rural craftsmen were hard hit by scarcity, hunger, death, sale of possessions, and loss of self-esteem. So were the millions of peasants living at the margin of self-sufficiency. Farmers with a surplus to be sold were better off as food prices soared. The Dutch farmers as a class fared reasonably well in the years of crisis. During the discussion on the 'Bill to encourage the importation of food commodities' the Hon. Hoffman mentioned the '... farmer, who finds in the high prices of his grain, more than compensation for the failure of his potatoes'[303]. The Governor of Groningen reported[304] 'The past year [1845] has been, on the whole, very advantageous to the agriculture in this province, not so much through a rich and abundant harvest, as through the high prices of all products'. The large semi-industrial estates in E Germany did exceedingly well, providing the needy cities (e.g. Berlin), exporting grain overseas, and selling potatoes or grain to distilleries rather than to the folks nearby.

Where cash was scarce on the countryside usurers took their chance. Many farmers became indebted and were, eventually, evicted. Social relations between labourers, farmers, and landowners came under stress. The situation was particularly serious (or well studied?) in France where the structure of rural society was uprooted and changed forever (Houssel, 1976: 242/4).

3.6. Conclusions[305]

1. The invasion of the European Continent (N of the Alpine Ranges and into W Russia) by *Phytophthora infestans* in 1844/5 had a tremendous impact because of the crop failure it caused, with famine and pestilence in its wake.
2. On the European Continent the harvest failures of 1845 and 1846, due to late blight on potatoes, rust on rye, voles, frost and drought, caused scarcity, hunger, and famine, leading to a recession with a drop in the demand of industrial products and a subsequent loss of employment, thus exacerbating the existent poverty and discontent.
3. Continental Europe suffered badly from crop failures in 1845 and 1846, leading to dearth, hoarding and speculation, rising prices, and finally resulting in an excess loss of lives tentatively estimated at ~700,000, a number yet uncertain but approaching that of Ireland.
4. Emigration overseas from Continental Europe seems to be in the order of one or a few hundred thousand persons, far less than the Irish emigration. Transmigration, from the countryside to the cities and the industrial areas, may have exceeded the transmigration of the Irish to England (~300,000) by far.
5. In Continental Europe existed a substratum of discontent, related to widespread rural pauperism and incipient urban proletariat, economic depression and empty treasuries, loss of trust, value loss of shares and real estate, bankruptcies, social unrest, and new political thinking. A general recession, with causes varying per state, came on top of the widely felt discontent and prepared the ground for the 1848 revolutions
6. The course of events in Continental Europe differed from that in Ireland. Ireland lost about one quarter of its population by death and emigration, but its political situation hardly changed. On the Continent the 1848 revolutions reshaped the political landscape, now determined by national Constitutions.
7. The 'potato murrain' was not the immediate cause of the revolutions, its pernicious consequences being rather the 'straw that broke the camel's back'. The epidemics of potato late blight, 1845 and 1846, together with many other agricultural misfortunes in those two years, contributed in an indirect way, but strongly, to the great political changes in Continental Europe, 1848.

Appendix 3.1. Contemporaneous information as published in The Official Newspaper of the Dutch Government (D: *Staatscourant*) of 1845.

'Disease' = potato late blight. Expected or estimated yields are given in fractions of normal yield, arranged per country or area and per date (day-month-issue):

Austrian Empire
23-10-251 Trieste – no disease.
10-11-266 Austria – disease present in Galicia, Vorarlberg, Upper and Lower Austria.
19-11-274 Galicia – disease progressed, most of harvest rotting, stoarability low.

Belgium
19-09-222 Harvest largely lost. Export of grain and potatoes forbidden.
01-11-259 No seed potatoes available, purchases in Scotland.

Denmark
28-10-255 Disease extending on late varieties – exports suspended.
05-11-262 Hardly any healthy lot available, cheap delivery to distilleries, exports suspended.
08-11-265 Rendsburg [presently Germany] – harvest ½.
25-11-253 Holstein – harvest ¼.

The Germanies
17-10-246 Bavaria hopes for normal harvest.
28-10-255 Bavaria – exportation forbidden.
17-10-246 Bremen – harvest 2/3.
28-10-255 Frankenland – yield good, rarely diseased.
08-11-265 Frankfurt [am Main] – harvest adequate but disease in storage.
17-10-246 Hannover – little surplus for export.
23-10-251 Königsbergen [Kaliningrad] – harvest good but disease present, ship loads lost by rotting[306].
03-12-286 Königsbergen [Kaliningrad] – harvest abundant, disease incidental. Storage rot, potatoes unfit for shipping.
03-12-286 Lithuania – harvest good.
03-12-286 Mazuria [Mazowze in present Poland] – harvest good.
01-11-259 Mecklenburg, Wismar – harvest <1/2.
18-12-299 Oldenburg – harvest 1/3, storability very poor.
23-10-251 Prussia – destruction.
01-11-259 Stettin [Szczecin, present N Poland] – harvest abundant but generally diseased.
28-10-255 Württemberg – exportation forbidden.

France
27-11-281 General overview – in 36 out of 86 departments harvest failure.
28-10-255 Alsace – harvest 2/3 to 3/4, rotting in storage.
05-11-262 Brittany, Brest – crops diseased – seed potatoes should be imported.
23-10-251 Brittany, Le Havre – harvest 1/4.
01-11-259 La Rochelle – harvest abundant and free from disease.
28-10-255 Les Landes – poor harvest, sometimes diseased.
23-10-251 Loire Valley – expected harvest 2/3.
28-10-255 Pyrenees– poor harvest, sometimes diseased.

Norway
01-11-259 Bergen – just enough for home consumption, no exports.
08-11-265 Norway – crop lost.

Russia
01-11-259 Memel [Klajpeda in present Lithuania] – harvest pending, no disease, yield moderate.
04-12-287 Russia – Exportation from Livonia [present Estonia and Latvia] and Kurland [Kurseme, present Latvia] forbidden.

Sweden
05-11-262 Disease near Göteborg, export forbidden.
08-11-265 Sweden – harvest was average.

Switzerland
28-10-255 Yield about 1/2, delivery to distilleries forbidden, prices low because of poor storability.

4. Crop loss in the Netherlands during World War I, 1914-1918: productivity of major food crops in a long-term perspective

During World War I, 1914-1918, the Netherlands were politically neutral, but they felt pressures from two sides, being dependent on the Allies for the incoming ships with grain and on the Central Forces for coal. Before the war, the Netherlands were a net exporter of food commodities. During the war Dutch agriculture had to adjust to self-sufficiency. Food scarcity was common, certainly among the townspeople. Though there was no acute hunger, under-nourishment was general, certainly among the workers' class. Food riots occurred and those of 1917 in Amsterdam were notable. This chapter discusses the productivity of the major food crops, potato, rye and wheat, in a long term perspective, with special attention for the effect of plant disease.

'Wij hebben de medewerking van Engeland en Duitsland nodig. De medewerking schijnt gekocht te moeten worden. Gekocht met aardappelen, die wij liever zelven zouden houden.'
We need the cooperation of England and Germany. It seems necessary to buy that cooperation. To buy it with potatoes that we would prefer to keep ourselves.

Mayor of Amsterdam. Gemeenteblad van Amsterdam 1917 Vol II-2 p1479.

'Als ik m'n man dat [= rijst] voorzet, krijg ik op me donder.'
If I put that [= rice] before my husband, I'll get hell

Protesting womens' comment on the offer of cheap rice. Gemeenteblad van Amsterdam, 1917 Vol II-2: 1460.

4.1. Potato riots, 1917

A policeman with drawn sabre stood guard at the front door of my grandfather's home in Amsterdam, 1917, during the potato riots[307]. At the time, my grandfather – Dr N.M. Josephus Jitta – , an amiable and peace-loving medical doctor, was one of the Aldermen of Amsterdam. His portfolio included public health and, during World War I, food rationing[308]. The riots began in a working-class district called 'Jordaan', a populous area housing many unemployed labourers with their large and hungry families. The immediate cause triggering the riots was the unloading of a ship with a cargo of old potatoes in the 'Prinsengracht', a canal bordering the area, the potatoes being intended for the military garrison. This rather tactless move triggered the fury of Jordaan women, of old a rebellious lot, who tried to loot the ship on 28 June, 1917 (Huijboom, 1992).

During World War I (1914-1918)[309] food was rationed in the Netherlands[310] as in most European countries. What caused the scarcity of food, and of potatoes in special, in a country that used to be an exporter of agricultural commodities, including potatoes? Were plant diseases involved in crop failure, as in neighbouring Germany in 1916 (Chapter 5)?

The present chapter investigates the effect of plant diseases on three major food crops and, indirectly, to assess the impact of plant diseases on national well-being during the war. The paper focuses on starch crops, potato, and, for purpose of comparison, wheat and rye. The Dutch phytopathological literature is remarkably silent on the topic of food supply in war time. Trade journals, such as *De Veldbode* and the *Overijsselsch Landbouwblad*, hardly mentioned potato late blight or other plant diseases[311]. The major source of information on the subject is the series of Annual Reports[312] published by the Department of Agriculture in the Ministry of Public Works, Commerce and Industry (§4.5.1).

4.2. Materials and methods

Integration levels. Data were available at various integration levels. The area and yield data for the Annual Reports were collected by the municipal authorities to be combined at the provincial and national levels. Yields were not determined scientifically but estimated. Generally speaking, farmers know and knew yields (and prices) quite well, but whether intentional over- and under-estimation may have occurred, I don't know.

Total loss of a crop at the field or farm level was a familiar and not infrequent experience, far more so than in recent times, when total crop failure is really exceptional. Reasons for total crop failure mentioned during the period of interest were damage to winter cereals by freezing, excess water, and slugs. Severely damaged fall-sown crops had to be ploughed in before the spring plantings (mainly in the north and west of the country). Night frosts in May-June could destroy potato and rye crops (mainly on the light soils of the east). Late blight could ravage the potatoes (mainly in the south-west) to the degree that crops had to be abandoned.

Many mishaps were experienced at the field or farm level but weather-induced problems usually affect larger areas, polders, municipalities, provinces, or the country as a whole. Drought, extreme cold, excess rain water, fresh water inundations, and inundations by sea water due to storm floods typically affect large areas covering several communities. In 1916, the Netherlands were hit by the storm flood of 13-14 January that inundated some 14,000 ha with salt water[313]. Since most of that inundated area was pasture land our data were not affected by this flood.

At the provincial level the large scale effects, the subject of the present study, became apparent, especially so when there were typical differences between provinces. At the national level long term trends were discerned (e.g. land use, productivity) and typical differences between years could be demonstrated (due to e.g. wet or dry, and cold or warm seasons).

Data selection and preparation. This chapter focuses on the major food crops, potato, wheat and rye, with emphasis on potato. To see more detail, two provinces per crop were compared, one in the north or north-east and one in the south or south-west of the Netherlands (Table 4.1). For cultural conditions, weather and diseases I followed the Annual Reports.

The potato categories 'ware', 'industry' and 'seed' were taken together, as only post-war data separated ware and industry potato. Seed potatoes were not mentioned separately. Similarly, data on autumn-sown and spring-sown wheat or rye were bulked. Where possible, the provincial and national yield data of potatoes were averaged over categories using the respective cultivation areas as weights. Spring wheat and spring rye usually covered less than ten per cent of the total areas under wheat and rye.

Table 4.1. The provinces chosen for comparison and some of their characteristics (seasons 1915-1918). Source: Annual Reports.

Province	Crops in order of of importance	Main soil types	Approximate hectareage in units of 1000 ha		
			Potato	Wheat	Rye
Groningen	Potato, rye	Clay, peaty sand	23	10	15
Friesland	Potato	Clay, sand	16	1	3
Zeeland	Wheat, potato	Clay	13	15	3
Noord Brabant	Rye, potato	Sand, clay	23	4	47

To study the effects of weather and disease on yield the statistic of interest was the yield per ha, expressed in the traditional way as hectolitre per hectare (hl/ha). The deviation of a yield in a particular year from the long-term trend was visualised by means of simple regression analysis. To reveal trends in productivity I used twelve seasons before and after the war, the periods 1903-1914 and 1919-1930, in addition to the war seasons 1915-1918.

The Annual Reports contain monthly means of daily temperature in °C, precipitation in mm per month, number of rain days (days with a precipitation ≥1 mm) per month, measured by the Royal Netherlands Meteorological Institute located near the centre of the country (Table 4.2). Weather data were used as illustrations only. Where appropriate the monthly figures were rendered as deviations d from the means over 28 years (1903 to 1930).

The summer weather was characterised by means of a 'Summer Index' (IJnsen, 1976), an index of temperature only. Figure 4.1A shows that the summer of 1917 was relatively warm. The other war summers were normal or 'at the cool side'. The winter weather was captured in a single figure, the 'Frost Index' (IJnsen, 1981). It only summarises winter temperatures. The winter 1916/7 (1917 in Figure 4.1B) was relatively cold, but the other war winters were mild. In crop loss studies precipitation is as important as temperature. Relevant precipitation data are given in Appendix 4.1.

Crop loss. Selected data sets were subjected to regression analysis of variable x_t (usually hl/ha) on time t (in years) over $n = 28$ years. The 'expected yield' (y_e) of a province in a particular year was that year's value on the regression line. The 'observed yield' (y_o) showed a deviation d[314] from the expected yield with $d = y_o - y_e$.

Data were normalised by calculating the 'standard error of the estimate' (S_x) for each data set. A dimension-less value δ is obtained dividing d by S_x ($\delta = d / S_x$). The value (δ) expresses the deviation of the observed from the expected value in a figure varying from ~-2 to ~+2. The 'approximate Wilk-Shapiro test' for normality of the distribution of δ over years showed no significant deviations from normality (Shapiro & Francia, 1972). This procedure permits a normalised comparison between two yield data sets (x_p and x_q) from different crops or provinces. To pinpoint interesting years the annual deviation Δ was calculated for the difference between δ_{xp} and δ_{xq}. A year was considered interesting when the absolute value of Δ exceeded 1, $|\Delta| > 1$.

Table 4.2. Monthly weather data averaged over 28 years, 1903 to 1930, as measured by the Royal Netherlands Meteorological Institute at De Bilt, in the centre of the country. Source: Annual Reports.

Month	Response		
	Temp. °C[1]	Prec. mm[2]	Rain days[3]
January	2.5	58.4	12.1
February	3.2	43.2	9.8
March	5.8	49.6	10.8
April	9.1	49.8	10.3
May	14.4	52.2	9.7
June	16.5	66.3	10.7
July	18.4	72.2	10.1
August	17.6	82.2	12.5
September	14.7	61.6	10.4
October	10.3	71.5	11.6
November	5.4	63.7	12.2
December	3.3	71.3	14.1
Year	10.1	742.5	134.4

[1] Temp. = Monthly mean of mean daily temperatures in °C; year = mean of monthly means.
[2] Prec. = Monthly precipitation in mm; year = mean annual precipitation in mm.
[3] Rain days = Monthly number of rain days (days with precipitation ≥1 mm); year = mean annual number of rain days.

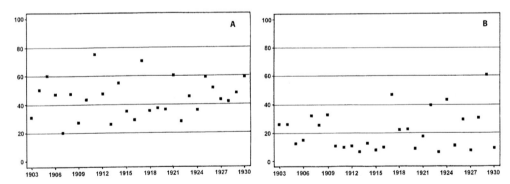

Figure 4.1 (A). Summer Indexes for the Netherlands over the years 1903 through 1930. The summer of 1917 was relatively warm. Data from IJnsen (1976). Horizontal – years. Vertical – Summer Index (the higher the index, the warmer the summer). (B). Frost Indexes for the Netherlands over the years 1903 through 1930. The winter 1917 (1916/7) was relatively cold. Data from IJnsen (1981). Horizontal – years. Vertical – Frost Index (the higher the index, the colder the winter).

'Crop loss' is defined as the difference between 'attainable yield' and 'actual yield' for any particular crop, year and area (Zadoks & Schein, 1979: 245). As the actual yield (y_o) of a province I used the 'observed yield' from the Annual Reports. For the attainable yield of a loss area (y_a) I took the 'observed yield' of a control province. To avoid systematic differences in yield level between provinces and years the observed yields were expressed as deviations δ from the expected yield, with δ_C and δ_L for the control and loss province, respectively. Crop loss in relative terms is $CL_r = \delta_C - \delta_L$.

Estimated crop loss (CL_e) in hl ha^{-1} is found by multiplying Δ with the S_y of the loss province. Estimated crop loss in per cent $(CL_\%)$ is calculated by relating CL_e to the expected yield of the loss area (y_e), $CL_\% = 100 \times (CL_e / y_e)$.

4.3. Results

4.3.1. National level – cultivation area

Potato was grown on both clay and sandy soils. The national potato area increased slightly but significantly (Table 4.3) from 1903 till 1930, though with considerable variation. Price expectations were a major cause of variability but adverse conditions may have had an effect, devastating night frosts foremost. The war years showed relatively large areas (Figure 4.2) under potato.

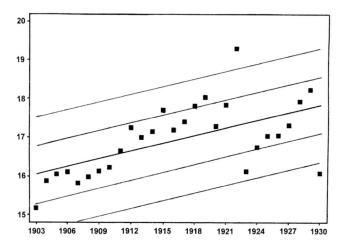

Figure 4.2. The national potato area (hectares) over the years 1903 to 1930. The area includes ware, industry and seed potatoes.
Horizontal – years. Vertical – area in units of 10,000 ha. The trend is shown as the regression line of hectareage on years, with additional lines at ± 1 and ± 2 S_y. The regression with slope 659 (ha/year) is significant (p = 0.0009, n = 28, r^2_{adj} = 0.32). Original data from Annual Reports.

Table 4.3. Summary table of national cultivation areas (ha), yield levels (hl/ha) and hectolitre weights (kg/hl) for wheat, rye and potato, the Netherlands, 1903-1930. Source: Annual Reports.

Year	Wheat[1]			Rye[2]			Potato	
	ha	hl/ha	kg/hl[3]	ha	hl/ha	kg/hl[3]	ha[4]	hl/ha[5]
1903	52,141	27.0	73.3	215,698	22.6	69.1	151,617	171
1904	54,081	28.8	78.7	212,849	22.0	73.0	158,732	209
1905	58,895	29.3	76.2	216,556	22.1	72.0	160,526	191
1906	53,730	30.7	76.7	218,220	22.5	71.7	161,114	209
1907	47,523	34.5	76.0	218,398	23.1	72.1	158,223	210
1908	51,320	32.1	-	219,566	25.2	-	159,887	213
1909	44,573	28.1	72.0	223,973	27.7	67.5	161,259	212
1910	54,784	28.6	-	219,956	24.4	-	162,215	192
1911	57,539	33.7	78.0	223,291	25.2	74.0	166,385	219
1912	57,854	34.1	72.0	228,044	24.9	70.0	172,344	249
1913	52,903	31.8	76.0	226,095	26.1	70.0	169,998	227
1914	56,500	33.9	76.5	227,674	20.9	69.5	171,513	248
1915	65,902	37.9	75.8	221,084	25.7	70.7	177,074	229
1916	52,438	31.1	74.0	199,855	20.5	69.7	171,833	216
1917	48,824	28.5	72.5	189,140	24.7	69.2	173,899	251
1918	60,009	31.9	75.0	191,165	24.0	71.7	177,952	258
1919	67,954	30.4	75.1	201,107	25.8	70.0	180,249	244
1920	61,489	34.3	74.3	199,270	25.7	69.4	172,884	248
1921	72,983	41.3	-	202,072	31.4	-	178,371	212
1922	60,551	35.8	-	202,168	29.9	-	192,884	297
1923	62,183	35.2	-	210,128	24.4	-	161,107	234
1924	47,898	34.6	-	197,903	28.1	-	167,456	247
1925	53,453	37.4	-	200,704	29.2	-	170,333	276
1926	53,340	36.8	-	197,298	24.7	-	170,336	260
1927	61,869	35.6	-	197,247	24.5	-	172,938	216
1928	59,943	42.8	-	196,135	31.6	-	179,103	322
1929	45,435	43.1	-	197,346	33.2	-	182,220	335
1930	57,518	37.7	-	192,374	27.7	-	160,712	281

[1] Data adding winter and spring wheat. Spring wheat covered at most 10% of area.
[2] Data adding winter and spring rye. Spring rye covered at most 10% of area.
[3] No data on 1908 and 1910; no data after 1920.
[4] Data adding ware and industry potatoes.
[5] Data averaged over ware and industry potatoes weighted by areas. Industry potatoes far out-yield ware potatoes but until 1920 no distinction was made. In Groningen, 1922, industry potatoes covered 61% of the area, out-yielding ware potatoes by 20%.

Wheat was grown on clay soils. The national wheat area showed great variation over the years without significant trend. A major cause of variation was the frequent failure of the fall-sown crops due to winter killing. Poor planting conditions in the autumn could affect the area sown. The ploughing under of failed winter crops was mentioned frequently and the failed crop was

On the political economy of plant disease epidemics

not always replaced by spring wheat. The expected price had some effect on planting decisions. The war years had relatively high wheat areas, 1917 excepted.

Rye was grown primarily on sandy soils. The national rye area was on the increase before 1915, dropped considerably during the war years, recovered briefly and declined. The war-time drop may have been induced by two factors. One was the lack of fertilisers that might have made rye growing unprofitable on the poor sandy soils, the other was to grow alternative crops with higher profitability. The decline after 1923 was probably due to improved soil and water management allowing substitution by crops with higher returns.

4.3.2. National level – yield in hl/ha

The upward trends in yield (hl/ha) of potato, wheat and rye over 28 years were highly significant (Table 4.4, Figure 4.3A), a well-established fact, explained by improved soil and water management, increased application of fertilisers, and plant breeding. Crop protection in the period 1903 to 1930 was yet of little importance in cereals, weeding excepted, but it may have played a role in potato, where late blight (*Phytophthora infestans*) could be controlled by application of Bordeaux mixture.

The upward trend in potato yields from 1903 to 1930 amounted to 3.6 hectolitre/ha/year. Variation in potato yields was considerable. Some of the outliers can be explained, others not. Interregional seed potato exchange occurred already in 1878 (Van der Zaag, 1999: 95), when Friesland exported seed potatoes to other Dutch provinces. Potato seed certification after field selection was practiced in Friesland around 1910 at a limited scale (Anonymous, 1910). Late blight control with Bordeaux mixture contributed to improved yields (various Annual Reports).

Table 4.4. Trends in potato, wheat and rye yields for selected provinces and for the Netherlands, 1903 to 1930. The trend is given as the linear regression equation for yield (hl/ha) on time (years). n = 28. Original data from Annual Reports.

Crop	Area	Constant[1]	p	Slope[2]	p[3]	r^2_{adj}	S_y[4]
Potato	Netherlands	179	0.0000	3.60	0.0000	0.58	24.2
Wheat	Netherlands	27.3	0.0000	0.395	0.0000	0.55	2.87
Rye	Netherlands	21.5	0.0000	0.250	NS	0.38	2.48
Potato	Friesland	232	0.0000	2.21	0.005	0.24	29.9
	Zeeland	132	0.0000	4.52	0.0001	0.44	40.3
Wheat	Groningen	32.2	0.0000	0.348	0.001	0.32	3.96
	Zeeland	29.6	0.0000	0.438	0.0000	0.48	3.74
Rye	Groningen	28.3	0.0000	0.364	0.0003	0.38	3.67
	Noord Brabant	20.9	0.0000	0.094	NS	0.06	2.41

[1] In hl/ha. Regression was calculated for (year – 1900). Thus 1915 was taken as 15.
[2] In hl/ha/year.
[3] $p > 0.05$ was considered non-significant (NS).
[4] S_y = 'Standard error of the estimate', used to draw the confidence limits at ± 1 and ± 2 S_y.

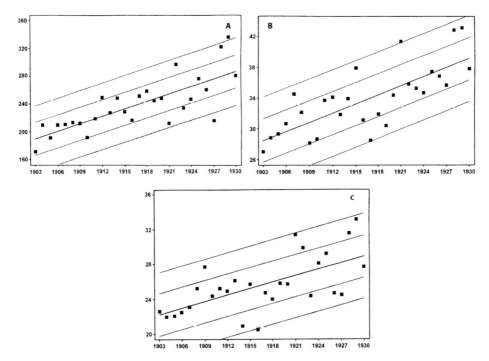

Figure 4.3. The national yields (hl/ha) of (A) potato[1], (B) wheat[2], and (C) rye[3] over the years 1903 to 1930. Horizontal – years. Vertical – yield in hl/ha. The trend is shown as the regression line of yield on years, with additional lines at ±1 and ±2 S_y. Original data from Annual Reports.
[1] The data include ware, industry and seed potatoes. The regression line with slope ~3.6 hl/ha is significant (p < 0.0000, n = 28, r^2_{adj} = 0.58).
[2] Taking winter and spring wheats together. Spring wheat did not exceed 10 per cent of the total area under wheat. The regression line with slope ~0.4 hl ha[-1] is significant (p < 0.0000, n = 28, r^2_{adj} = 0.55).
[3] Taking winter and spring rye together. Spring rye did not exceed 10 per cent of the total area under wheat. The slope of the regression line is non-significant.

National wheat and rye yields (hl/ha) showed a similar upward trends (0.4 hl/ha/year for wheat and 0.25 hl/ha/year for rye; Figures 4.3B and 4.3C), again with considerable scatter. Some of the outliers can be explained. Wheat and rye yields were highly correlated. The correlation is due primarily to the general upward trend in yield, but its relatively high r^2_{adj} (= 0.63) suggests that the two crops were subject to the same ecological factors, see e.g. the high of 1915 relative to the low of 1916, or the peak of 1928 relative to the trough of 1927. The point was also apparent from some typical comments in the Annual Reports (Appendix 4.1).

The weather data of the Meteorological Institute, far outside the major cropping areas, and the comments of the Annual Reports did not always correspond. In the period 1903-1930 the wettest year was 1903, with plenty late blight. The driest year was 1921, at the middle of a set of three dry years. The coldest year was 1917. Warm summers occurred in 1911, 1913, 1917, 1921 and 1930 (Appendix 4.1). From a meteorological point of view the war years 1915 to 1918 were not exceptional. The years 1915 and 1916 had a high precipitation whereas 1917, beginning with a cold and long winter, had a hot June month.

4.4. Crops and their major harmful agents

4.4.1. Potato

Harmful agents. According to Quanjer[315] (1921a) two diseases dominated the European potato scene, late blight (*P. infestans*) and 'degeneration'. The Annual Reports mentioned two other harmful 'agents', night frosts and late summer and autumn rains. Night frosts were frequent in late May and early June on the sandy soils in the SE (e.g. 1905). Late summer and autumn rains favoured late blight and hampered timely lifting, with lots of rotting tubers and poor storage quality as the results (e.g. 1930).

Late blight. Wet seasons favoured the sporulation, dispersal and infection of *P. infestans* (e.g. 1903). Wet falls furthered the overwintering of *P. infestans* in groundkeepers, potato refuse, and seed tubers. Varieties differed in resistance to late blight but these differences are ignored here. As a rule, early varieties suffered less than late varieties[316]. Since Millardet (Schneiderhan, 1933) potato late blight could be controlled by the application of Bordeaux mixture and this was increasingly and successfully done in the Netherlands (Figure 4.4)[317]. According to the Annual Reports two applications sufficed. In the Netherlands hand-operated back-pack sprayers were utilised and later hand-held line sprayers with a central reservoir and pump[318]. In 1904 the Netherlands counted 606 horse-drawn 'spraying machines'[319]. Later counts were not available. Spraying against late blight decreased during World War I because copper became scarce and expensive[320]. Unfortunately little detail is available so that the effect of decreased spraying cannot be assessed.

Figure 4.4. Spraying equipment (Quanjer, 1911).
Knapsack sprayer made by Vermorel (France), Model 'Éclair N° 1'. 14.5 liter. Price DGL 21.
Horse-drawn sprayer, with six nozzles, made by Besnard (France), Model 'Dispositif'. Price of set-up equipment without cart DGL 115. In real life the driver sits on the cart.

Degeneration had been known since the 18th century (§5.3.2). Its symptoms were described as 'curl'. The classical remedy was two-fold. First, new and not-yet-infested varieties could be grown from seed[321]. Second, planting material could be obtained from areas with relatively little infestation such as Scotland (England), Friesland Province (Netherlands), Eiffel Mountains (Western part of Germany), or selected fields (Eastern part of Germany). Friesland as a seed potato growing area had already some fame around 1910[322]. 'Green lifting' of plants for seed potatoes was an old practice (e.g. Kops, 1810: 357), explained nowadays as avoidance of tuber infection.

The cause of 'degeneration' was unknown and hardly studied. It was associated with 'curl', mentioned frequently in the Annual Reports. Quanjer (1921b) stated that 'curl', the typical field symptom of degeneration, was in fact a disease syndrome from which he isolated three separate diseases for detailed studies, 'leaf roll' (phloem necrosis), 'mosaic' and 'crinckle'. Salaman (1949) attributed the 'curl' to Potato Virus Y. The typical 'leaf roll', due to the leaf roll virus, caused a first large-scale outbreak in Germany, 1905[323]. Yield loss could be up to 80 per cent (De Bokx, 1982). Wennink (1918) measured losses of over 90 per cent due to leaf roll. Westerdijk (1916) described the damage potential of potato mosaic (30 to 60 per cent yield loss).

Oortwijn Botjes (1920) demonstrated aphid transmission of leaf roll and mosaic, Quanjer *et al.* (1916) had shown transmission by grafting. Quanjer (1921b) thought of a sub-microscopic parasite but he avoided the word 'virus'. The plant scientists at the time were not yet ready to accept viruses as plant pathogens (Westerdijk, 1917) though, according to the standards of the time, adequate proof was given in the Netherlands for 'mosaic' in tobacco (Mayer, 1882a,b) and in tomato (Westerdijk, 1910).

Areas. Two provinces were selected for comparison, Friesland in the N and Zeeland in the SW (data not shown). The Frisian data included up to ~10 per cent of industry potato. Zeeland did not grow industry potato. The Frisian area under potato showed great scatter, with considerable decrease after 1922, whereas the area in Zeeland increased irregularly but significantly. During the war years 1915 to 1918 the area under potato had a dip in 1916, especially in Zeeland, a dip attributed to the ploughing under of many potato fields destroyed by late blight. Relatively large areas were planted in Friesland after 1916, with a peak in 1918.

Yields. Trends in potato yields from Friesland and Zeeland were compared (Figure 4.5, Table 4.4). The S_y of Zeeland by far exceeded that of Friesland, suggesting that yield stability over the years in Zeeland was much lower than in Friesland. Zeeland with its lack of fresh water may have been more sensitive to summer drought, hampering potato development, than Friesland. In some years the δ values of the two provinces were remarkably close (1910, 1920, 1921, 1925), whereas in other years the δ values differed considerably, with Δ = 2.27 in 1903 and Δ = 2.38 in 1929 (Table 4.5). In 1903, the wettest year out of 28, late blight ravaged the north of the country. During 1929, the one but driest year, potatoes in Zeeland must have suffered from drought. The effects of these two harmful agents on yield at the provincial level was of the same order of magnitude, |Δ| > 2, or a yield deficit of over 60 hl/ha.

Comments. National yield levels during the war years did not differ much from the expected values, with an exceedingly bad yield in Zeeland, 1916 (a wet year), and a peak yield in Friesland, 1917 (a nice summer). In the post-war year 1919 potatoes suffered from leaf roll, mosaic, super mild mosaic, and *Verticillium* wilt. A cold spring caused a hollow stand. Lack of fertilisers led

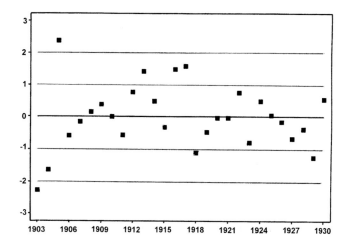

Figure 4.5. Potato yields compared between Friesland and Zeeland. Scatter plot of the difference Δ between the Friesland (x_o) and Zeeland (x_e) values of the relative deviation δ of the observed from the expected yield. Horizontal – years. Vertical – Δ. Only values of Δ ≥ |1| are used for further interpretation. For explanation see text.

to modest yields on sandy soils. Early freezing in the fall damaged potatoes in the field and in the pit (Annual Report on 1919: xvi-xvii). Usually we could not single out definite causes for specific provincial losses, but Table 4.5 gives some indications.

4.4.2. Wheat

Harmful agents. Among the physical agents often mentioned were winter freezing, either severe frost or repeated freezing and thawing, and excess winter wetness, leading to poor soil conditions for the rest of the season. Among the biotic agents were slugs in the fall and winter, voles, and rust. This rust was probably yellow stripe rust (*Puccinia striiformis*; Zadoks, 1961). Summer storms with heavy rain could cause lodging. Rainy summers often delayed the wheat harvest and caused loss of quantity by shedding and/or sprouting in the ear, and loss of quality by sprouting in the ear and rotting on the field or in the stack[324]. The complex of foliar diseases which drew so much attention during the second half of the 20th century, among which the septorias[325], was never mentioned in the Annual Reports over the period 1903 to 1930. Occasionally mention was made of bunt (*Tilletia caries* in Zeeland, 1923).

Areas. Two typical wheat provinces, Groningen in the NE and Zeeland in the SW of the country, are climatically different, Zeeland having the milder climate. Harvests in Groningen are about 2 weeks later than in Zeeland. Weather fronts sweep over the country from SW to NE within a day so that a rain period, post-harvest in Zeeland, may be pre-harvest in Groningen, causing considerable damage to the Groningen harvest due to sprouting in the ear and rotting in the stook or stack, as in 1927 (Table 4.5). Cold air invading from the NE may cause serious winter damage in Groningen but not in Zeeland, as in 1907 and 1908 (Appendix 4.1).

Table 4.5. Loss estimates (hl/ha) of potato, rye and wheat using pairs of provinces with a yield difference exceeding one standard error of the estimate, S_y with possible causes of loss.

Crop	Year	Control area[1]	Loss area[1]	δ_C [2]	δ_L [3]	Δ [4]	S_y [5]	Loss (hl/ha)	Possible causes
Potato	1903	S	F	0.18	-2.09	2.27	29.9	68	Late blight in the north
	1904	Z	F	1.19	-0.46	1.65	29.9	49	Poor soils in the north
	1905	F	Z	0.50	-1.88	2.38	40.3	96	Late blight Zeeland
	1913	F	Z	0.71	-0.71	1.42	40.3	57	Late blight Zeeland?
	1916	F	Z	-0.28	-1.77	1.49	40.3	60	Curl, late blight Zeeland
	1917	F	Z	1.45	-0.14	1.59	40.3	64	
	1918	Z	F	0.76	-0.36	1.12	29.9	33	Late blight Friesland
	1929	Z	F	2.26	1.00	1.26	29.9	38	
Rye	1909	N	G	1.31	0.28	1.03	3.67	3.8	
	1911	N	G	0.86	-0.22	1.08	3.67	4.0	
	1916	G	N	-0.80	-2.82	2.02	2.41	4.9	
	1923	N	G	-0.36	-1.55	1.91	3.67	7.0	
	1927	N	G	-0.89	-1.91	1.02	3.67	3.7	Sprouting in the ear
	1928	G	N	2.35	0.65	1.70	2.41	4.1	
	1930	N	G	0.20	-1.29	1.49	3.67	5.5	
Wheat	1905	G	Z	1.35	-0.80	2.15	3.74	8.5	Summer storms
	1910	Z	G	0.91	-0.32	1.23	3.96	4.9	Poor wet soils in NE
	1912	G	Z	1.61	0.53	1.08	3.74	4.0	
	1915	Z	G	-0.81	-1.83	1.02	3.96	4.0	Good weather in SW
	1917	G	Z	0.86	-0.41	1.27	3.74	4.7	Winter freezing Zeeland
	1927	Z	G	-0.43	-1.56	1.13	3.96	4.5	Sprouting in the ear?
	1929	Z	G	2.30	1.24	1.06	3.96	4.2	
	1930	G	Z	0.12	-1.13	1.25	3.74	4.7	Sprouting in the ear?

[1] N = Noord Brabant, F = Friesland, G = Groningen, Z = Zeeland.
[2] δ_C = The deviation of the observed from the expected value, control area.
[3] δ_L = The deviation of the observed from the expected value, loss area.
[4] $\Delta = \delta_C - \delta_L$.
[5] S_y = 'Standard error of the estimate', used to draw the confidence limits at ± 1 and ± 2 S_y.

We only considered those hectareages that deviated more than one S_y from the trend line. The levels of the two trend lines differ significantly at $p<0.001$ but their difference in slope is not significant. Both provinces had low hectareages in 1913, with no obvious reason, and in 1924 and 1929, due to winter freezing. High hectareages occurred in 1921, possibly a post-war effect. Zeeland had relatively low areas in 1906 and 1910, Groningen in 1903. The respective areas during the war years 1915 to 1918 did not deviate much from the trend line, though Groningen had a relatively high area in 1915 and a low area in 1917, the latter due to winter freezing.

Yields. Again, we only considered yields (hl/ha) deviating from the trend line by more than one S_y. High yields in either province occurred in 1921, 1928 and 1929, mainly due to favourable summer weather. Low yields in both provinces occurred in 1919, when harvest conditions were miserable. Obvious differences between the two provinces are shown in Table 4.5, with tentative explanations.

In some years the δ values of Groningen and Zeeland were remarkably close (1906, 1928), whereas in other years the δ values differed considerably, with $\Delta = 1.27$ in 1917 and $\Delta = 2.15$ in 1905. Such large differences indicate specific regional effects within a relatively homogeneous country (Table 4.5). In 1917 frost damage was considerable, especially in Zeeland. In 1905 wheat in Zeeland may have been hit by summer storms, rain and hail.

Comments. During the war years 1915 to 1918 six out of the eight regional wheat yields considered were within the range $\delta < \pm 1\ S_x$, that is not far from their expected values. The two exceptions were in Zeeland, with one very good year (1915) and one very bad year (1917). In 1916 the harvest weather was miserable, leading to relatively low yields in either province. In 1918 yields were moderate.

4.4.3. Rye

Harmful agents. Damage by winter cold and by summer rains was comparable to that in wheat. At flowering and early seed set rye was susceptible to night frosts, late May and early June, as often mentioned in the Annual Reports. Lodging due to heavy summer rains occurred incidentally. Rust, probably *Puccinia recondita*, and nematodes, probably *Ditylenchus dipsaci*, were mentioned occasionally. During the war, when labour was scarce, weeds became damaging, especially black grass (foxtail, *Alopecurus myosuroides*).

Areas. Two rye-growing provinces were compared, Groningen in the NE and Brabant in the S. In Brabant the decline of the rye area began before the war. The rye area reached a low level in 1918, and stayed at an even lower level after the war. In Groningen the rye area was at its lowest in 1917, with a remarkable peak just after the war.

Yields. Brabant yields did not increase significantly over the years 1903 to 1930 but in Groningen, technically more advanced, they increased by 0.36 hl/ha per year, on average (Table 4.4). The scatter in the Groningen rye yields increased markedly over the study period. Groningen and Brabant had about equal δ's in 1906, 1907, 1920 and 1925. Low yields occurred in both provinces, in 1914, 1918, 1923, 1927, and 1930. For possible explanations of these low yields see Appendix 4.1. In Groningen yields were low in 1916 and 1917 and in Brabant in 1926. Years with relatively high yields were the dry years 1921 and 1929 in these two provinces, 1928 in Groningen and 1909 and 1922 in Brabant. Between Groningen and Brabant Δ varied, with values of 1.91 in 1923 and 2.02 in 1916 (Table 4.5). The causes of the differences are not obvious.

Comment. The war years 1915 to 1918 show fair yields in 1915 and rather poor yields in the other years but in Brabant, 1916, the yield was disastrously low. The scarcity of fertilisers, mentioned in the Annual Reports, was not clearly reflected by the harvest figures in the two provinces considered, but it may have influenced the decision not to plant rye in Brabant, 1918.

4.5. Discussion

4.5.1. The data

All data used here are derived from the Annual Reports. Their weakness is in the aggregation of field and farmer data to municipal data. Occasionally, the Annual Reports mention alternative data pathways yielding somewhat different results. We may question the correctness of the data received by the municipal officers and the reliability of the data transmitted by them[326] to the higher levels, but as these data have been collected for nearly a century in more or less the same way by over 900 municipalities, they are thought to be acceptable for purposes of comparison, certainly at the higher integration levels.

Statistics on cultivation area were available. Cultivation areas were subjected to the discipline of the market and, in times of dearth, to the regulation of the government. These aspects are not discussed here. Sometimes, the adversity of the weather and, in its sequel, the damage by disease affected recorded cultivation areas. Abandonment of crops because of winter killing, night frosts, and late blight was mentioned above and in Appendix 4.1.

This paper concentrates on yield data expressed in hl/ha. Weights in kg/hl were available for wheat and rye up to 1920, but not for potato (Table 4.3). Though hectolitre weights are indicative of quality their use did not provide interesting new information.

4.5.2. Crop loss

As specification of yields in hl/ha was current during the period under consideration, I calculated crop loss in terms of hl/ha, which is adequate for the present purpose, the identification of good and bad years. These are identified by δ_c and δ_L being either both positive (potato, 1929; Table 4.5), or both negative (wheat, 1915; Table 4.5). If the available information is pertinent on the cause of the loss one may conclude that the loss due to that cause is 'up to' or 'in the order of' the calculated value.

In addition to the quantitative losses estimated here the growers suffered qualitative losses (storage, processing, and consumer qualities) that may have been financially more important than loss of quantity. The Annual Reports sometimes alluded to these qualitative losses without providing sufficient information to make any calculation.

4.5.3. Causes and effects

Many factors may cause damage but only a few, well-known factors cause such large-scale damage that their effects become visible in aggregated data, at the provincial or national level. For the period 1903 to 1930 some factors can be identified with reasonable certainty (Appendix 4.1, Table 4.5).

A wet autumn may delay the potato harvest, favour late blight and bacterial rot, and produce seed potatoes of poor quality as in 1909, leading to low yields in 1910. In addition, a wet autumn may delay or preclude the planting of winter wheat as in 1923, reducing the 1924 wheat area considerably (Table 4.3).

On the political economy of plant disease epidemics

A 'bad winter', bad for agriculture, may be too wet, too cold, or both. On water-soaked soils winter wheat may suffer from open stands due to slugs and mice (1903), or to alternate freezing and thawing that damages the rootlets. Poor soil conditions may lead to low yields (1910, Groningen). Severe frosts cause winter killing and sometimes total loss of winter wheat (1907, 1909, 1917, 1924, and 1929) and winter rye (1917, 1924).

In late spring and early summer (May, June) night frosts may injure potato and rye crops, as the Annual Reports often mentioned, but in the aggregated data the effects are not obvious, rye in 1914 excepted. Heavy rainfall in late spring and early summer may lead to weed and lodging damage as in wheat and rye (1913) but in the aggregate the effect is not clear (rye in Groningen, 1923, excepted).

Summer drought reduces potato growth and leads to low yields (1921) but such a drought may be followed by a rich potato harvest in the next year[327] (1922). Summer rains at the time of the grain harvest can be very damaging to grain quality and even grain quantity by kernel loss, sprouting in the ear, or rotting in the sheaf, stook or stack (wheat in Groningen, 1919 and 1927, and in Zeeland, 1930; rye in Groningen, 1927, but not in Brabant). Heavy rains in late summer may trigger severe late blight on potato (1903, 1905, and 1927). In 1927 the rains were so heavy as to cause failure of Bordeaux mixture applications. In 1903 potato suffered much late blight in Friesland but not in Zeeland and in 1905 the obverse happened (Table 4.5). When the rains continued well into the fall potato lifting[328] was hampered and tubers rotted in the soil or in the pile.

4.5.4. World War I and Dutch Agricultural policy

With the tacit consent of the major belligerents, Germany and Great Britain, the Netherlands remained neutral during World War I (Frey, 1998: 13). Neutrality required careful manoeuvring by the Dutch Government to safeguard crucial imports and exports.

At the time, Dutch agriculture was part of an open economy, as it is today, with imports of basic commodities, food, feed and fertiliser, and exports of a variety of agricultural end-products. Among the imports (Figure 4.6) were wheat from Russia and the USA for human consumption, maize from the USA as animal feed, and fertilisers such as potassium from Germany, nitrogen from Chile and phosphorus and sulphur from England and Belgium. Among the exports were dairy products, fruits, vegetables, fresh potatoes and various potato products, among which potato flour. Exports to Germany continued well into 1916[329], when political pressure by England brought food exports to Germany to a stop[330]. Siney (1957: 248) stated that the British interfered 'in every part of Dutch life'.

In the summer of 1914, just before harvest time, the able-bodied young men were enlisted in the services and a great many horses were requisitioned for the not-yet-motorised Dutch army. Fortunately, harvest problems due to lack of labour or draught power had been avoided thanks to the combination of fine harvest weather and unexpectedly great mutual assistance among the farmers[331]. When war was definitive, imports decreased rapidly but exports continued for some period. The government took appropriate measures to provide for the availability and equitable distribution of fertilisers[332] but in the course of the war seasons 1915 to 1918 the lack of fertilisers was increasingly felt. Dutch agriculture was certainly hampered by lack of labour, draught power (horses), fertilisers, and copper sulphate. Most references to these

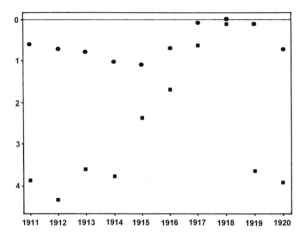

Figure 4.6. Dutch wheat and rye imports just before, during, and just after World War I.
Horizontal – years. Vertical - Net imports of wheat ■ and rye ● in units of 100,000 tonnes. The 1916 imports of wheat were ~7.5 x 10^5 tonnes per year or ~300 grams per person per day. Source: Annual Report over 1920: Table XXXIX.

shortages in the Annual Reports are incidental. At the provincial and national levels the effects of these shortages on yield figures are not evident.

The government supervised the transition from an open export-oriented to a self-contained, autarchic economy (Sneller, 1943: 110) by means of a growing set of injunctions, price control measures, and centralised purchasing of food commodities to be redistributed among the population at large[333] (Jansma & Schroor, 1987: 288/9). Food exports came to a stand-still[334] (Figure 4.7).

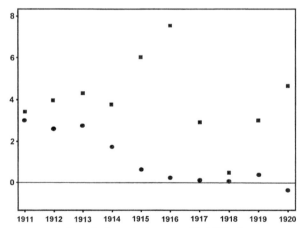

Figure 4.7. Dutch potato exports before, during, and just after World War I.
Horizontal – years. Vertical - Net export of fresh potatoes ■ and of potato flour[335] ● in units of 100,000 tonnes. The 1915 exports of fresh potatoes amounted to ~2.4x10^5 tonnes per year or ~100 grams per person per day. Source: Annual Report over 1920 Table XXXIX.

Table 4.6 resumes the differences between observed and expected yields for potato, wheat and rye of the war years 1915 to 1918. The expected yields are the yields predicted by the trend lines. The calculated differences (hl/ha) are expressed in terms of the 'standard error of the estimate' δ. When δ is positive the observed yield exceeds the expected yield. When $\delta \geqslant 1$, the deviation is considered interesting. When $\delta \geqslant 2$ the deviation is statistically significant at $p \leqslant 0.05$ and considered to be important (rye, 1916).

Table 4.7 shows the total production of starch crops in the Netherlands during World War I. Using 1916 yields and a population of 6.5 million people, refugees included, and estimating the proportion for seed potatoes and storage and transportation losses at 10%, the availability per person per year is about 5 hl or 350 kg. The 1910 consumption was about 250 kg potato per person per year (Bieleman & Van Otterloo, 2000: 238), so that the potato supply apparently was adequate, at least in principle. For rye the availability is about 0.56 hl or 39 kg and for wheat 0.33 hl or 16 kg. The 1910 consumption was 100 kg rye and 100 kg wheat per person per year. Hence, the rye and wheat supplies were quite deficient. They were supplemented by stored and imported grain, at least in the early war years. There was plenty of rice in store but the working class refused to eat it[336]. 'Government bread' contained an admixture of at least 5% potato flour, making it unsightly and unpalatable to many.

Consumers had to compete with alternative uses such as animal consumption (hogs), potato flour (farina) production, and exports. The rioters, or those behind them, objected indignantly to these exports (§4.1). The Dutch government, trying to observe strict neutrality, wanted at least some exportation to maintain good relations with the belligerent parties[337]. Ships and trains loaded with potatoes were looted in the ports of Amsterdam and Rotterdam. The real problem, however, was the provision with cereals of which the national production was, roughly, in the order of ~20 per cent of the human home consumption. Imports continued throughout the war period but did not suffice to feed animal stocks.

Table 4.6. Starch crop yields in the war years 1915 to 1918.
Entries δ are deviations from the respective trend lines expressed in standard errors of the estimate (S_y). Note that δ is mostly positive in 1915, a 'good' year, and negative in 1916, a 'bad' year. Source: Annual Reports.

Area	Crop	Observed yields (hl/ha)				δ per harvest year			
		1915	1916	1917	1918	1915	1916	1917	1918
Netherlands	Potato	229	216	251	258	-0.17	-0.85	+0.45	+0.59
	Rye	25.7	20.5	24.7	24.0	+0.18	-2.02	-0.42	-0.81
	Wheat	37.9	31.1	28.5	31.9	+1.63	-0.88	-1.92	-0.88
Friesland	Potato	298	259	313	261	+1.10	-0.28	+1.45	-0.36
Zeeland	Potato	257	133	203	244	+1.42	-1.77	-0.14	+0.76
Groningen	Rye	35.0	31.2	32.7	31.1	+0.34	-0.80	-0.49	-1.02
Noord Brabant	Rye	24.3	15.6	19.5	21.3	+0.83	-2.82	-1.24	-0.54
Groningen	Wheat	39.5	34.2	34.9	36.7	+0.53	-0.90	-0.81	-0.45
Zeeland	Wheat	42.2	33.4	30.2	34.9	+1.61	-0.86	-1.83	-0.69

Table 4.7. Total production (hl) of starch crops in the Netherlands during World War I. Source: Annual Reports.

	Potato	Rye	Wheat
1915	44,663,432	5,679,084	2,498,337
1916	37,051,037	4,103,766	1,686,573
1917	43,689,588	4,673,071	1,391,744
1918	45,913,166	4,589,070	1,913,895

The war years were not particularly good crop years but they were not as unfavourable as in Germany (Chapter 5). Only the season of 1916 had rather low yields of potato, wheat and rye. This may have been due in part to lack of fertilisers but that suggestion is not supported by the yield data from 1917 and 1918. During the very cold winter of 1917 an unknown but possibly considerable amount of potatoes was frozen in transit or storage in railway cars, and lost for consumption.

The Dutch population (1914-1918: about 6.5 million[338]) was maybe somewhat undernourished[339], but it survived the war period without great problems even though it was temporarily enlarged by nearly one million Belgian war refugees[340]. Protests and demonstrations against food shortages typically occurred in early summer, when the stores had been (nearly) emptied and the standing crop had not yet been harvested. Demonstrations took place in many towns in the early summer of 1916[341]. In 1917, riots occurred in several cities[342]. The Amsterdam riots were controlled with difficulty, at the expense of 114 wounded and nine casualties[343]. The upheavals were spontaneous and, apparently, unexpected. A major objection was the continued exportation of potatoes that the Dutch Government deemed necessary to please the belligerent parties. After the riots the Dutch Government decided to send an extra train-load of potatoes to Amsterdam.

The question remains to what degree the townspeople went hungry. The *Gezondheidsraad* (National Health Council) stated that public health was not suffering much from the food scarcity. Some Amsterdam physicians made an enquiry[344] under labourers' families (Sajet & Polak, 1916). They reported a contrary opinion as they concluded that the diet of these families was marginal at best and, certainly for the children, inadequate in the long run. Under-nutrition was not yet acute, but resistance against diseases was undermined and tuberculosis was a threat. The authors stated that the real paupers were far worse off. This was 1916, when the worst was yet to come. The problem clearly had two sides. One was the limited supply of food, the other the lack of purchasing power due to war-induced unemployment of the labourers. In addition, there was the refusal to eat unfamiliar food (rice).

The initially unfriendly attitude of the townspeople toward the growers in agriculture and horticulture was moderated in 1916 by a better understanding[345]. The obverse happened in Germany[346]. During the war, the Germans suffered far more from food shortages than the Dutch. In Germany losses due to weather and plant diseases contributed decisively to the population's distress.

4.6. Conclusions

1. This paper opened and ended with the 'potato riots', Amsterdam, 1917. The inhabitants of Amsterdam suffered from food shortages. The poor were under-nourished but they did not starve.
2. Nearly one million Belgian refugees, sojourning temporarily in the Netherlands, 1914/5, could be fed adequately[347].
3. The Dutch Government kept the Ship of State afloat with a relatively modest set of injunctions and regulations but it could not avoid all food riots or protests.
4. Dutch agriculture managed to adjust to war conditions and to remain productive notwithstanding lack of labour, draught power, fertilisers, and pesticides.
5. Areas of starch crops fluctuated under the influence of the weather and of profitability considerations.
6. Yields of starch crops varied considerably but were fair in view of the long term trend, with the exception of 1916 when yields were low.
7. Plant diseases (and pests) took their share as usual. They could be a definite nuisance at the field and farm level. Potato losses in Zeeland, 1916 and 1917, were at least 60 hl/ha.
8. Seen in a long-term perspective (1903-1930) the negative effect of pests and diseases on productivity of Dutch agriculture during World War I was not excessive.

Appendix 4.1. Overview by years, 1903-1930.

Overview of weather conditions and crop damages, extracted from a large body of data in the 'Annual Reports' (Department of Agriculture in the Ministry of Public Works, Commerce and Industry), and annotated.

Year = harvest year. Some meteorological data are given by way of illustration. The *d* stands for 'deviation from the 28-years mean (Table 4.2), with between brackets the period of *d*, one or a few months, or a year. The next letter indicates the response involved, *d*P = deviation of the monthly precipitation in mm, *d*R = deviation of the monthly number of rain days, *d*T = deviation of temperature in degree-months (degrees as °C).

S = season, 1 = winter (January – March), 2 = spring (April – June) 3 = summer (July – September), 4 = autumn (including potato harvests and winter wheat planting; October – December).

Weather, crop and disease notes – Brief abstracts from the originals by the present author.

Year	S	Weather, crop and disease notes
1903; *d*P(Apr) = + 76.7; *d*P(year) = + 184.4	1	In 1902 difficult autumn planting. Winter 1903 rather warm. Nonetheless *winter wheat* damaged by frost, mice & slugs. In Groningen small area harvested, in some polders 75 % ploughed, remainder open stands.
	2	Spring fairly cold. Excess water in April.
	3	Wet summer. *Winter wheat* – low yields. *Rye* – good yields but quality loss. Harvest difficult.
	4	*Potato* with heavy late blight, in Groningen many fields not harvested, in North Holland many diseased tubers. Yields low. Autumn wet, poor tillage and sowing. Slug damage in *winter wheat*.
1904	1	Slug damage in wheat, much winter wheat ploughed in Groningen and North Brabant.
	2	Climatically favourable season. March and April dry. *Spring wheats* rusted[348].
	3	*Wheat* – open stands but fair yields of good quality. July through September dry. *Rye* severely damaged by rye nematode in Limburg, worst in 25 years.
	4	*Potato* had only little late blight. Wet fall.
1905	1	Winter normal.
	2	March warm, April cold. Night frost (24 May) damage in *potato* in S and severe in *rye* in E.
	3	June with hail. Good year for *cereals,* Zeeland excepted. Regional rain fall differences[349].
	4	*Potato* good in Friesland, poor in SW due to late blight beginning July[350].

1906	1	12-03 Storm flood, 3600 ha lost in Zeeland.
	2	Normal.
	3	*Wheat* yield and quality fairly good, SW excepted. *Rye* generally good. Groningen, severe hail damage June 1st, torrential rains in July damaged all crops.
	4	*Potato* – yield good and quality excellent, though curl[351] in S and much late blight in SW. Very low yields in Zeeland. Autumn relatively dry and favourable.
1907; dT(Jan+Feb) = -2.7; dP(year) = -106.5	1	Severe frost in January with damage to cereals. *Wheat* ploughed under in Groningen[352].
	2	Cold. *Rye* suffered from night frost. Much *spring wheat* sown.
	3	Dark, cool, wet. *Wheat* with peak yields, specially *spring wheat*.
	4	*Potato* good, with little late blight but severe curl[353]. Nice fall. Plantings late.
1908; dT(Jan) = -3.5; dP(year) = -126.5	1	Severe frost alternating with thawing. In Groningen 20% of *winter wheat* ploughed under.
	2	Severe night frost April 25th. May with heavy rain and hail damages.
	3	*Rye* very good but locally in E hail, wireworms, ergot. Locally summer rain damage[354]. *Potato* – no night frost, little late blight, some curl[355].
	4	*Potato* good, little late blight and curl. Fall good for tillage and planting.
1909; dT(Jan+Feb+Mar) = -5.5	1	Winter with severe frosts. Much *winter wheat* ploughed under in N[356].
	2	June cold and dry.
	3	July and August with heavy rains. *Wheat* yields low, with much sprouting in the ear. *Rye* yields low[357]. Potato curl prominent in some areas[358].
	4	Fall mild, very wet. *Potato* yields satisfactory but lifting late.
1910	1	Mild but wet winter with inundations in river district. Soils in poor condition.
	2	Favourable. Emergence of *potatoes* irregular due to bad quality of tubers.
	3	*Wheat* with low grain and straw yields due to poor soil condition. *Rye* harvested in bad weather.
	4	*Potato* suffered from poor quality of seed potatoes; low yields. Bordeaux mixture had good effect. Yields low. Crop failed in Friesland due to late blight and curl. Nice fall weather.
1911; dP(Jly) = -39.2; dP(year) = -99.5	1	Mild.
	2	April cold and dry. *Potato* suffered from night frost in NE.
	3	Dry, especially in July. *Rye* good, *wheat* very good.
	4	No late blight. Very good *potato* harvest. Tillage and sowing good.
1912; dP (Aug) = +84.8; dP (year) = +171.5	1	Favourable winter with only a few days severe frost in N.
	2	Spring planted crops irregular and thin.
	3	Cold, wet, late. From August onwards much rain. *Rye* yields good[359]. Wheat yield and quality poor when harvested after summer rains.
	4	*Potato* harvest late with high yields, on clay soils in W many rotten tubers. Much wetness. Poor tillage. Plantings delayed.
1913; dP(May+Jne) = +44.5	1	Mild winter.
	2	Rains in May & June caused severe weed & lodging damage.
	3	In E severe damage by white grubs (larvae of *Tipula* spp). *Wheat* in Groningen had much whiteheads[360]. *Rye* good. Early, untreated *potato* varieties killed by late blight[361].
	4	*Potato* profited from spraying. Fall dry, favourable for planting

1914	1	Mild.
	2	Groningen: severe rust in *winter barley*. Feet and trousers coloured orange-yellow (probably *Puccinia hordei*). Wet March and April with dry E winds lead to crust formation. *Wheat* in Groningen grassy, rust[362] disquieting.
	3	May with severe night frosts (*rye*). Much rust[363] in *rye*. Summer fair. Harvest normal notwithstanding mobilisation.
	4	*Potato* gave rich harvest of good quality. Good autumn.
1915;	1	Mild, very wet.
dP(Jly+Aug) = +52.6;	2	May & June with rains and night frosts. *Potato* in sandy areas suffered
dR(Jly) = + 4.9		from 3 night frosts, losses in NE up to 80%.
	3	Late July wet, hail damage; late August wet, *Wheat* – good yields, modest quality; wet harvest in Groningen.
	4	*Potato* with little blight in S, but much in Friesland due to lack of labour and of copper for spraying. Nice autumn.
1916;	1	Wet, little frost, *winter wheat* damaged by water-logging. Zuyderzee dikes
dP(Dec+Jan+Feb) =		broken due to storm flood.
+101.1;	2	March & April wet. *Potato* area reduced because of poor financial results in
dT(Jne) = - 2.5		1915 and good prospects of sugar beet, onions and other crops
	3	June cold, late August wet. *Barley* had much rust[364] in early summer, low yields. *Wheat* harvest wet, yield and quality modest, hl weight low. *Rye* suffered from poor weather, weediness, lack of labour & lack of fertiliser.
	4	*Potato* yields low by lack of fertiliser & labour, much weediness[365], heavy curl, much late blight. Fall favourable, but tillage and sowing late.
1917;	1	Severe winter January – March, many *winter cereals* lost.
dT(Jan+Feb) = -6.7;	2	April cold, May clear & dry. Rye retarded, meagre due to lack of fertilisers.
dP(Oct) = + 82.6	3	June clear & dry. *Potato* in N damaged by night frost 6/7 June. *Winter wheat* and *winter rye* with low yields due to winter frost damage. *Potato* rarely treated because of expensive copper but yield and quality generally good.
	4	October wet. *Potato* and beet harvests delayed by rains. Late blight late, damage only in low lying areas of South Holland. Fall too wet, few winter crops sown.
1918;	1	Mild.
dT(Jne) = -1.5;	2	Early spring, May cold and dry.
dP(May+Jne) = -57.5	3	June cold and dry, night frosts 5-7 June. *Wheat* yield and quality good except Groningen. *Rye* with much lodging, otherwise yields good. *Potato* with poor foliage due to cool and dry spring, and lack of fertiliser; top necrosis, curl, rhizoctonia, Verticillium wilt, but little late blight[366].
	4	Wet, *potato* lifting late, many rotten tubers.
1919;	1	No problems.
dP(year) = -47.5	2	Rather cold, wet March. Late, poor spring tillage. May sunny and dry with 4 frost nights.
	3	June & July cool & wet, good *wheat* and *rye* harvests, Groningen excepted. *Potato* in Zeeland of poor stand, in Friesland with late blight. *Potato* yields moderate, much curl mosaic, top necrosis, and Verticillium wilt.
	4	Dry fall, early planting. November frosts.

1920;	1	Mild winter, wet till February.
dP(year) = -126.5	2	February and March dry, early tillage. *Winter wheat* with good stand.
	3	June sunny, July sufficient rain, August cold & wet, good *cereal crops*.
	4	*Potato* fair which rather much though variable late blight, in France frozen in the pits. Dry fall.
1921;	1	Dry, good winter crops.
dP(year) = -306.5	2	April too dry, crops delayed.
	3	Dry, *winter cereals* good yields of good quality. Rye yields exceptionally high.
	4	Ware potatoes good, but chain tuberisation. No late blight.
1922;	1	Alternate freezing & thawing. *Wheat* and *rye* partly ploughed under. *Rye* with poor stand.
dT(Jan+Feb) = -2.7	2	March & April bleak with nightfrosts. May dry & warm, much spring wheat sown.
	3	June & July rainy, August cold. *Rye* recovered but straw short.
	4	*Potato* yields peaked (as in 1912 after dry year), prices low, much leaf roll & top necrosis. Relatively dry fall.
1923;	1	Mild, frost & snow, warm March.
dT(Jne) = -3.5;	2	April dry, May & June cold.
dT(Jly) = +2.6;	3	6-14 July hot. *Wheats* good, much straw, rust in Gelderland & bunt in
dP(Oct) = +50.6;		Zeeland. *Rye* with heavy crops, lodging.
dR(Oct) = +9.4	4	*Potato* regular. Very wet fall, lifting difficult. *Winter crops* sown late or not at all.
1924;	1	Cold & long winter, many *winter cereals* ploughed under.
dR(Aug+Sep) = + 9.1	2	Favourable May, nightfrosts 5 & 6 June.
	3	June & July good, August & September wet, poor harvest weather.
	4	*Potato* suffered from sudden late blight in July with incidental losses in the SW 25 to 75 %. Fall normal.
1925;	1	Mild.
dP(Dec) = +69.7	2	Dry March, little night frost.
	3	June – August dry, about 1 August many thunderstorms, tornado on 10 August, September wet, good year for *cereals & potatoes*.
	4	Fall-sown crops damaged by inundations from heavy December rains.
1926	1	Frost damage, dry March.
	2	Cold May.
	3	June through September favourable. *Wheat* and *rye* – much straw but meager yield.
	4	Favourable. *Potatoes* locally killed by late blight[367].
1927;	1	Mild.
dP(Jne) = +69.7;	2	April wet & cool, May dry with night frosts.
dP(Aug+Sep) = 62.1	3	June very wet, August & Sept wet. *Rye* with much sprouting in the ear.
	4	*Potato* suddenly with late blight, spraying had little effect due to heavy rains, much rotting in storage, prices up. Favourable autumn. December with snow & frost.
1928	1	Frost damage in March.
	2	April & May cold, May wet.
	3	June cool & sunny, July and August sunny, high *grain* yields.
	4	*Potato* with much 'blue' but no late blight. Autumn good for tillage and planting.

1929;	1	Severe & long winter, *wheat* in N lost, in SW protected by snow.
dT(Jan+Feb) = -11.2;	2	Favourable. *Rye* with open stands, remedied by extra N fertiliser.
dP(Year) = -151.2	3	Dry and sunny, *wheat* and *rye* yields high.
	4	Very large *potato* harvest, no late blight, prices down. Favourable autumn.
1930;	1	Mild. No winter damage.
dP(Year) = +130.5	2	Dry, rains in May.
	3	Very hot days in June. July, August & September rainy. *Cereals* with wind damage, much lodging, sprouting in the ear, rotting.
	4	*Potato* harvest late, much late blight, difficult to control on soaked soils, modest quality, much refuse.

5. Crop loss in Germany during World War I, 1914-1918: productivity of major food crops and the outcome of the war

Starving the civilian population was one of the instruments of war during World War I. In this aspect the Allies were more successful than the Central Forces led by Germany. Germany had bad luck as the weather during the war was generally unfavourable to agriculture. Mobilisation of men and horses, combined with shortage of manure and fertilisers, reduced agricultural production. Lack of human care enhanced plant diseases and pests. Plant diseases and pests, in turn, greatly reduced yields of the main starch crops. Some 700,000 Germans died of hunger and corollary disease. The fighting spirit of the German troops suffered greatly by the misery at home and the poor food at the front. The highly developed, science-based German agriculture failed to sustain the warring nation. The main culprit, however, was neither weather nor plant disease but the complete disorganisation of food production, food storage and food distribution, due to the utter neglect of the nation's food base by the military dreaming only of a rapid victory that did not materialise.

'In the happiness of the subjects lies the happiness of the king and in what is beneficial to the subjects his own benefit. What is dear to himself is not beneficial to the king, but what is dear to the subjects is beneficial (to him).'

Kauţilīya arthaśāstra, Part II: 1.19.34 (Kangle, 1986).

'Da dies alles fehlte, setzte mit Ausbruch des Krieges eine völlige Kopflosigkeit ein.'
As all this lacked a complete brainlessness set in with the outbreak of the war.

F. Aereboe (1927: 48) on agricultural policy and stocks of grain in 'Der Einfluss des Krieges auf die landwirtschaftliche Produktion in Deutschland'.

5.1. War and plant disease

War and plant disease form an unpopular combination largely neglected by the phytopathological literature. There are exceptions. Vanderplank (1963) had the courage to write a chapter 'Plant disease in biological warfare', a topical theme in the present period with anxiety of bioterrorism. Horsfall & Cowling (1978) discussed the role of plant disease epidemics in human history under two headings, 'Impact of plant disease epidemics on war' and 'Impact of war on plant disease epidemics'. One case was the blow that potato late blight dealt to Germany during the first World War, 1914-1918. Apparently, they built their case on data provided by Carefoot & Sprott (1967: 87), but that story[368] painted only part of the picture.

Indeed, crop loss due to late blight during World War I contributed to Germany's misery, but there is more about it. This chapter is a non-exhaustive study of the crop protection problems, and of some agricultural problems in general, of a country at war. It focuses on the major starch crops, potato, rye, and wheat. The easily accessible German phytopathological literature led to only a few specific (e.g. potato late blight) and non-specific (e.g. 'degeneration' of potato) diseases, and to several more general problems such as drought and voles. Supposedly, other diseases such as 'blackleg' of potatoes were important but they were mentioned rarely in the papers consulted.

The historical epidemiologist is interested in disentangling the factors that led to an epidemic, both environmental and human factors, supposing that there is one (and only one) clearly defined epidemic with one identifiable pathogen on one host crop. He might also be interested in the consequences of such an epidemic on the human condition, expressed in conduct, distress, or warfare. However, historical reality defies the natural scientist's bias to reduce complex situations to a one-to-one relationship. Accepting this challenge, I will try to disentangle 'the impact of war on plant disease epidemics' and 'the impact of plant disease epidemics on war' in the case of Germany during World War I.

5.2. The war situation

5.2.1. General position

During World War I, 1914 to 1918[369], two parties on the European scene attempted to starve each other. One party, Great Britain, belonged to the 'Allies' with, among others, France, Russia and, as of 1917, the United States of America. The other party, the 'Central Powers', was formed primarily by imperial Germany and the Austrian-Hungarian double monarchy. War fronts existed in western and eastern Europe, the Balkan, the Far East, the Near East, and Africa.

The war was not only a world war[370] because of the many fighting arenas, but also because it was explicitly directed towards the non-fighting civilian population by means of bombing open cities and of intentional starvation of civilians. To this purpose Britain blockaded Germany by closing the Channel and the North Sea, intercepting any overseas shipping to or from Germany. Germany tried to blockade Britain by means of its submarines in the high seas, cutting off its food imports. In the end, England was more successful than Germany.

Before World War I Germany was a constitutional monarchy, with a weak parliament and a strong monarch, the Emperor Wilhelm II. It was also a conglomerate of numerous principalities and a few free cities, all with their own jurisdictions. Foreign affairs and war were centralised functions of state. Geographically Germany covered the area of the present Federal Republic of Germany extended to the West with the Alsace and to the East with part of present Poland up to the river Wisła (G: *Weichsel*), and to the North-East even beyond (East Prussia). Population size (1913) was ~67 million persons (Meher, 1917).

Germany up to 1914 had a strongly protected, science-supported, highly productive agriculture[371]. Nonetheless it imported wheat from Russia (including Ukraine) and the Americas. Rye, produced in surplus, was exported 'in exchange of wheat' so to say.

Several other commodities had to be imported such as meat, fat and vegetable oils, all paid by the export of industrial products. Germany produced large surpluses of potato that were used as animal feed and as inputs for the alcohol and starch industries. The war changed this situation drastically. Food imports rapidly decreased. Germany tried to become self-supporting but, unfortunately, the war situation imposed many restrictions on German agriculture[372], among which lack of labour, draught power, manure, fertilisers, and machinery. Gradually food became scarce (Berthold, 1974).

Most able-bodied young men were called to arms[373]. The high wages in the war industry enticed many farm workers. With so much work to be done by women, children and elderly an amazing productivity was maintained. Lack of labour in agriculture is mentioned regularly (e.g. Fischer, 1915) but its effect seems limited since apparently the rural population had a surplus of labour. The necessary work was done though its quality may have suffered. The large enterprises in the East, employing hundred thousands of prisoners of war (Berthold, 1974), did better than the small farmers. The latter often ploughed and planted in wet soil, a practice[374] said to be disastrous for the potato crop.

Draught power (horses, oxen) was reduced quantitatively, due to requisition of over one million horses by the army, and qualitatively, due to lack of fodder and concentrates that decreased the effectiveness of the remaining draught animals. Tillage was neglected more and more, with increasing loss of soil fertility (Aereboe, 1927: 37). Weeds became an overwhelming problem so that children brigades had to go weeding[375]. In 1917 and 1918, the military made draught power available for field work in spring and autumn.

The number of livestock (horses, cattle, hogs) decreased gradually by slaughtering, thus reducing available manure. In addition, the amount and quality of manure decreased by lack of concentrates. Even green manure was deficient as the 1915 harvest of clover and serradella (*Ornithopus sativus*) seed failed and the seed became too expensive[376]. The supply of fertilisers decreased drastically. Potassium remained available and some phosphorus, but most nitrate and phosphorus came from guano (Chile saltpetre) that was no longer available. Soils impoverished in the course of the war years.

New machinery was not produced and spare parts of old machinery became scarce. Mechanics for the maintenance of agricultural machinery were in the army and machinery gradually broke down. Threshing was delayed by late delivery of fuel, lack of electricity, and absence of spare parts; only the field mice rejoiced. The war industry needed supplies otherwise used in agriculture, among which nitrate for explosives and copper for bullets.

5.2.2. Food policy

The military, convinced of their own superiority, expected the war to be of short duration[377], a few months only. No food had been stored in advance (Aereboe, 1927 p29). A rapid conquest of large territories would ensure Germany's food supply. This was a major miscalculation. A technical reason was the limited stock of explosives made from imported nitrate, guano primarily (Herzfeld, 1968: 259), a limitation removed by the timely development of the Haber-Bosch procedure[378], 1913, for the production of ammonia that was transformed into nitrate. Little nitrate was available to agriculture (Kielmansegg, 1980: 171). The carelessness of the military with respect to food security was such that the war department did not even employ one agricultural scientist.

When the years of reality replaced the months of expectation several policy measures were taken, often ineffective or even counter-productive, that need not be discussed here. The end result was a general decline of the agricultural production (Table 5.1) due to exhaustion of soils, poor tillage, neglect of crop care, and weediness. The area under major food crops decreased due to increased fallowing and shifts toward more profitable, non-starch crops. National yields of major food crops, cereals and potatoes, went down by one third.

Science-based agriculture flourished with large estates in the east (present Poland) whereas small farms dominated in the south and west. Though some minor efforts (see below) were made agricultural science was unable to support the war economy. German agriculture was unfortunate during the war years as it was plagued by adverse weather reducing the productivity of the crops directly, either by drought or excess rain, or indirectly, by diseases.

Table 5.1. Germany's production of rye, wheat and potato during the war years, 1914-1918.
At the time, Germany encompassed the Alsace and a sizeable part of present Poland. Total production in millions of metric tonnes (M t's) and in per cent of mean over 1908-1913, tonnes per hectare (t/ha), and Millions of hectares (M ha). Data sources: Flemming (1978: 84). The M ha values were estimated from total production and hectare yield. Note that Aereboe (1927: 86) and Berthold (1974: 92, 99, 101) stressed the unreliability of the agricultural statistics in Germany during the first decades of the 20th century. According to Flemming (1978: 81) harvest data were systematically 10-15 per cent too high. Nonetheless, the trend, as shown by the data presented here, is clear.

Year(s)	Rye				Wheat				Potato			
	M t's	%	t/ha	M ha	M t's	%	t/ha	M ha	M t's	%	t/ha	M ha
1908-13	11.2	100	1.8	6.2	4.1	100	2.1	2.0	45.9	100	13.8	3.3
1914	10.4	93	1.6	6.5	4.0	98	2.0	2.0	45.6	99	13.5	3.4
1915	9.2	82	1.4	6.6	3.9	95	1.9	2.1	54.0	118	15.1	3.6
1916	8.9	80	1.5	5.9	3.1	76	1.8	1.7	25.1	55	9.0	2.8
1917	7.0	63	1.3	5.4	2.3	56	1.5	1.6	34.9	76	13.7	2.5
1918	6.7	60	1.4	4.6	2.3	56	1.7	1.4	24.7	54	10.1	2.4

5.3. Agents of crop loss

5.3.1. Biotic and abiotic agents of loss

Agriculture had gradually learned to cope with all kinds of damaging agents causing crop loss. The agricultural science of the time aimed primarily at good crop husbandry. Selection of profitable varieties was ongoing but plant breeding was yet in its infancy. Crop protection as a discipline did already exist but was not yet very effective. Soil and water management were on their way. The phytopathological literature of the war years discussed only few pests and diseases in sufficient detail to be referenced here.

Agriculture could deal with abiotic agents such as extreme cold or heat, and excess or lack of water, up to a degree. When the excess was marked nothing could be done, as in the 1911 drought when yield loss of potato was over 20 per cent at national level (6 to 38 per cent at provincial level; Schander, 1917: 146). When fall-sown crops were frozen in winter, spring crops could be planted to reduce the losses.

The control of biotic agents, as far as possible, was primarily by means of good care, seed selection, mechanical or even hand-weeding, crop rotation, and on-farm hygiene. Chemical control was hardly existent, except against potato late blight and grape mildews. Chemical seed disinfection was emerging (Schmidt, 1917: 69). Usually, weather related outbreaks of pests and diseases could not be stopped.

An interesting feature is that several diseases considered important in recent times, among which various foliar diseases of wheat, did not receive mention at the time though they were already known. They may have been considered part of the normal farmer risk and not worthy of attention. It is difficult to believe that they were not present at all[379].

5.3.2. Potato – degeneration

'Degeneration' was known since the 18th century[380]. The classical remedy was two-fold. First, new and not-yet-infested genotypes could be grown from seed. Second, planting material could be obtained from areas with relatively limited infestations such as the Eiffel Mountains (Western part of Germany) and selected fields (Eastern part of Germany). In Germany, the produce from fields with little infestation could be sold as seed potatoes. Degeneration (G: *Abbau*) in potatoes was a major problem in Germany. Many studies were devoted to degeneration, also known as 'curl' (G: *Kräuselkrankheit*). Appel (1906a,b) recognised 'curl' as a complex from which he singled out a disease he called 'leaf roll' (G: *Blattrollkrankheit*), that had caused an epidemic in 1905[381].

Quanjer's pioneering work, ultimately leading to the distinction of a number of different virus diseases, was ignored or even rejected[382]. Quanjer (1921a) himself thought highly of German potato research with the exception of degeneration studies that he dubbed 'a hopeless failure' (D: *hopelooze mislukking, ibid.*: 13). Quanjer (1921b) stated that 'degeneration' was a complex syndrome from which he had already separated distinct diseases called 'leaf-roll', 'mosaic' and 'crinckle'. He had taken these diseases, tentatively attributed to ultra-microscopic parasites, in 'pure culture' (*in planta*, in living plants). In Germany, however, degeneration was seen as a physiological disorder, probably soil-induced[383], 'heritable' at best. The literature suggests a 'not-invented-here' attitude, leading to systematic disregard of foreign publications[384].

Degeneration was especially common on the heavy soils of the Rhineland Province[385]. Rhineland regularly imported seed potatoes from the Eiffel mountains, where healthy crops grew in a slightly cooler and more windy climate, and later from Eastern Germany, where good seed potatoes could be grown on light soils[386]. During the war the delivery of seed potatoes was suspended[387]. Thus, Rhineland had to use home-grown seed potatoes which were much degenerated. This degeneration became apparent as an increasingly open crop, with missing plants. Many surviving plants produced leafroll-like and blackleg-like symptoms. The symptomatology of leaf roll was well-established at the time (Appel, 1906a,b) but knowledge of other plant viruses, and of their identification and epidemiology, was nascent at best.

Schander (1917) quoted an experiment, apparently performed near Bromberg, Pommerania (now Bydgoszcz, Poland), with 'original' seed potatoes and their progeny. In the relatively good years

1912 and 1913 the yield loss due to degeneration was some 46% (average of 12 varieties) and in the war years 1915 and 1916 ~47% (average of 23 varieties). Schander (1917) also quoted an experiment in which three generations of seed potatoes are compared simultaneously in 1916 (Table 5.2). After two generation the loss was ~37 percent[388]. The 'year effect' was obvious.

Table 5.2. Yields of three generations of potato planted simultaneously in 1916. Data are means of 6 varieties. From Schander (1917).

Generation	Year of origin	Yield in tonnes/ha
Original seed	1915	21.0
First generation	1914	20.2
Second generation	1913	13.1

5.3.3. Potato – late blight

The predominant idea in Germany was that late blight should be controlled by means of resistance breeding[389] rather than by spraying Bordeaux mixture. Schander, well aware of the favourable effect of Bordeaux mixture[390], thought treatment useful for the early and mid-late varieties but not for the late varieties with their exuberant production of sprawling vines, then and there, that made penetration of the crop by men, horses and equipment impracticable. In the war years no treatments were applied because the necessary personnel and horses were not available. Pulverisation (Souheur, 1892) had not been successful by lack of a suitable compound. During the war the copper in the Bordeaux mixture was needed for bullet shells and this may have been another, unmentionable, reason not to use Bordeaux mixture[391] at a time when it would have been most appropriate to do so.

The suggestion was made (Quanjer *et al.*, 1920: 54; §3.2.5) that degeneration or curl made potato plants more susceptible to late blight[392]. Whether this is true or not, potato late blight was the second problem in German potato cultivation.

5.4. A crop loss chronicle

In this chronicle the data are arranged according to harvest year. Weather systems sweep over Europe north of the Alpine ranges and have similar effects in different countries. The prime phytopathological example is the outbreak and rapid spread of potato late blight in 1845 and the quite dissimilar epidemic of 1846 all over NW Europe, preceded and followed by the very cold winters 1844/5 and 1846/7 (Chapter 3).

Weather data were taken to be more or less uniform over Germany. This is a very rough approximation at best since Germany, 1914, stretched out over more than 14 degrees longitude and 7 degrees latitude, approximately one thousand km in E-W and 800 km in N-S direction. The simplification is justified by focussing on the pattern. Regional variations and localised events, super-imposed on the general weather patterns, are not considered here.

Per year some dominant features of the weather are given, followed by damages to the main starch crops, potato, wheat and rye (Figures 5.1A and B). A selection was made since only large effects are relevant to this story. Unfortunately, the crop protection literature dealing with the war period is not rich.

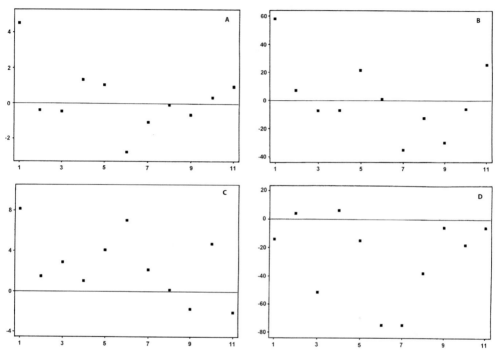

Figure 5.1. The 1916 weather in Bremen, NW Germany, as compared to a long-term average.
(A) Mean monthly temperature in °C in excess of the long term average. (B) Precipitation in mm per month in excess of the long term average. (C) Number of rain days per month in excess of the long term average. (D) Sunshine in hours per month in excess of the long term average. Original data: Ritter (1917), long-term average not specified.

5.4.1. 1914 – the rust year

Weather. The relatively mild winter 1913-14 allowed the over-wintering of yellow stripe rust and voles. Winter wetness and snow in March favoured *Fusarium* snow mould and foot rot in rye in the Rhine Province, Westfalia and Pommerania (Schmidt, 1917: 68/9). A dry spring caused a crusty soil and hence a delay of the spring-sown crops. The summer was warm (Zimmermann, 1916a).

Potato. The potatoes had relatively much late blight and bacterial rot (Anonymous, 1916), but the potato production in 1914 was about normal, with some 'blackleg'[393] in the Austrian areas (Kornauth, 1915).

Cereals. The wheat crop, harvested just before the outbreak of the war, suffered severe attacks by yellow stripe rust[394] (*Puccinia striiformis* Westend.). Rust infections extended from NE Netherlands through Germany, well into Austria and present Czech Republic and modern Poland[395], where it was exceptionally severe in Silesia, in the area of Bromberg (present Bydgoszcz, Poland). In Saxony, district Halle, the winter wheat suffered losses in the order of 20 per cent. In N Germany spring wheat was retarded and became subject to a yellow rust epidemic. The effect of disease on yield is not clear but, since imports of wheat were incidental at best, losses must have weighed heavily.

An outbreak of voles, probably with the common vole (*Microtus arvalis* Pall.) as the main culprit, occurred in 1914, sometimes in clover[396] but also in wheat, where it could be devastating. In 1914, voles were quite active in various parts of Europe, as in the Ukraine where some 11,000 ha of wheat was completely destroyed (Rossikow, 1916). An unusually severe outbreak occurred in W Prussia[397] and Poznań (present Poland), from spring to fall[398]. Control was possible in principle[399] but labour-intensive, and probably received little attention in 1914. The fall-sown rye in Mecklenburg suffered from vole and slug damage, and from a sudden November frost (Zimmermann, 1916b).

5.4.2. 1915 – the drought year

Weather. The early summer of 1915 was extremely dry, the drought causing great losses in cereals and some damage to potato. Mommsen (1995 p688) stated explicitly 'harvest failure', whereas Kielmansegg (1980: 1777) called the harvest 'not very good'. Grain harvests were bad in Hungary, 1915 (Hirschfeld *et al.*, 2003: 461).

Potato. The early summer of 1915 was very dry and the rains toward the end of June came too late for a good yield of the early varieties. The late varieties did extremely well. '*Gipfelrollen*' ('top roll') occurred frequently. Schander (1917: 154) describe 'top roll' elaborately without suggesting a causal agent. Probably he saw *Verticillium* wilt caused by *V. albo-atrum* and/or *V. dahliae*. Early and severe frosts occurred over large areas in N Germany on 21 September and following days. In the end, a bumper crop yielded at least twice the amount needed[400]. Possibly, it was a good year for aphids and virus spread[401], with some after-effects[402] in 1916.

Storage became the great problem (Löhr vom Wachendorf, 1954: 114/5). The government, having confiscated most of the produce, had taken the responsibility for storage but it had no appropriate storage facilities. Railway carriages were used to store the potatoes; many potatoes were frozen during the winter, and lost. Cellars of town halls and schools were filled with potatoes. These, stored at too high temperatures, gradually began to rot. In the spring of 1916 the stench became unbearable and the school kids had to stay at home with 'stench vacation' (G: *Stinkeferien*). The dirty mess had to be cleared, the precious food was gone, and the enormous surplus had evaporated by a 'nearly criminal waste'[403].

So far, I did not find independent information on these storage losses. Schander[404] provided indirect evidence when discussing the need of drying potatoes before storing them in cellars; he stated 'In the year 1916 [but preceding the 1916 harvest] hundreds of tonnes of potatoes [of the 1915 harvest] could have been saved and utilised for human consumption, when the drying had begun in time'. Probably, harvest time had been rainy, with tuber infection by late blight and bacterial pathogens. Early in 1916 farmers kept lots of potatoes in store to feed their animals (Kielmansegg, 1980: 178). Farmers, of course, knew how to store potatoes but the authorities in their superior wisdom followed their own fancy[405].

Cereals. The rye crops in Mecklenburg, some already damaged in the preceding fall, suffered from alternating freezing and thawing in March, so that serious harvest failures occurred (Zimmermann, 1916b). The wet winter of 1914-15 led to severe *Fusarium*[406] problems in Pommerania (present NW Poland). *Fusarium* was exceptionally frequent in Bavaria (Schmidt, 1917). Drought caused a loss of cereal production estimated at 20 to 25 per cent of the 'normal' yield. The drought during early summer caused many crop failures in cereals (Schander, 1917: 154).

5.4.3. 1916 – the blight year

Weather. Spring and early summer 1916 were cool and moist, with many rain-days, but without excessive precipitation. Ritter (1917) in Bremen gave a phenological description of the year 1916. Spring was exceptionally early (~14 days). In contrast, late summer was slightly retarded (~4 days), whereas the fall was very early again (~13 days). Temperatures were about normal except for a warm January and a cold June. May was wet and the months July through September were dry. The year was poor in sunshine with a deficit of 289 hours. The number of sunshine hours per month was very low in March and in the months May through August. The number of rain-days per month was above normal with peaks in January and June (Figure 5.1). The weather, generally wet and dull, had a serious impact on agricultural production. 1916 was a bad crop year in England, the Netherlands, and Germany[407].

Potato. A Potato Ordnance (G: *Kartoffelverordnung*) required the farmers to save as many tubers as possible for human consumption (G: *Saatgutsparnis*; Schander, 1917: 165). Several methods were available. (1) The archaic practice of cutting seed potatoes in two or more parts was revived in 1916. The practice was attractive since the tubers harvested from healthy 1915 crops were few but large. Under favourable, warm and not too wet conditions the practice did not cause problems but it was disastrous in the cold and wet spring of 1916, leading to many diseased and missing plants (Table 5.3). (2) Wide planting was often applied but caused yield reductions (Table 5.4). (3) Small seed tubers were planted, a practice without much risk of loss when the tubers are really ripe and disease-free, but very risky when they are not fully ripened and/or diseased. Method (1) was not frequently used and had little effect on national yield, but methods (2) and (3) were more common and may have contributed to the harvest failure[408].

May and June showed typical April weather, leading to slow and late emergence of the 1916 crop. Seed potatoes that were cut or blighted suffered most and often did not emerge at all. Crops remained open and tuber formation was delayed. These problems were more serious with early

Table 5.3. Schander's 1916 experiment with halved tubers versus intact tubers, the tuber halves and the intact tubers being equal in weight. From Schander (1917).

Tubers	Yield in tonnes/ha	Missing plants %	Yield loss %
Intact	24.4	2.5	-
Halved	15.1	22.5	38

Table 5.4. Schander's 1916 experiment at Kleschewo on planting distances. No additional specifics given. One Zentner per Morgen = 196.1 kg/ha. From Schander (1917).

Variety	Yield in Zentner/Morgen at planting distance		Loss
	Wide	Narrow	%
Wohltmann	132	117	11
Imperator	92	70	24
Ella	109	103	5
Industry	123	109	11
Mean	114	100	12

plantings and on heavy soils, late plantings faring much better. Later, frequent rain showers caused severe crusting of the soil under the open crops, notwithstanding regular ridging.

Schander (1917) discussed a special problem, the quality of seed potatoes for the 1916 planting. The early summer of 1915 had been excessively dry and the potato crops had shown specific symptoms, indicated as 'top roll' (G: *Gipfelrollen*). Varieties differed in drought sensitivity, the mid-late varieties suffering most. The late development of many 1915 potato crops, with poor ripening, supposedly had a negative effect on the quality of the seed potatoes planted in 1916 (Kuhn, 1918). Indeed, many seed potatoes were found to be diseased, leading to open stands. Holdefleisz (1917) added another thought. In Saxony, district Halle, he observed great differences between potato crops, even on one field or farm. Some crops hardly emerged. As large and apparently good seed tubers often produced miserable crops the variation was, again, attributed to seed potato quality. Holdefleisz suggested that the severe September night frosts, that killed many stands right-a-way in 1915, had damaged the tubers to the degree that they could hardly produce sprouts and shoots[409].

Remy (1916: 818), working in W Germany (Bonn), complained that the 'seed potato supply, so exceptionally difficult due to the war', had caused a nasty situation in the Rhine districts. Schander (1917: 150), stationed in the east[410], confirmed the existence of an 'export restriction', a regulation limiting the usual shipping of seed potatoes from east to west[411]. Remy thought that this official measure was the major cause of the potato harvest failure in W Germany, 1916, since the Rhine districts had been forced to use their own seed potatoes transmitting the feared degeneration (G: *Entarting, Abbau*). The clay soils of the west were known to be conducive to degeneration whereas the light soils of the east generally produced seed potatoes practically free from it[412]. Maybe, the early and severe drought of 1915 stimulated the multiplication of aphids and hence the transmission of virus[413], with effects becoming visible only in the next year, 1916.

The potato season 1916 was disastrous with potato late blight (*Phytophthora infestans* (Mont.) Berk.) as the most conspicuous cause of disaster. As most of the summer was cool and cloudy with a rainfall that was not excessive but regular (a high number of rain days), the conditions for a late blight epidemic were ideal. Warming up in July, 1916, the weather remained wet, creating perfect conditions for late blight and also for bacterial blackleg[414]. Early and mid-late crops were ruined and yields reduced to 50%, at times even to 25% of the expected value. In many fields the early and mid-late potatoes were not worth lifting. Sometimes, only nut-sized potatoes were produced[415]. Differences in disease severity occurred between locations, soils and varieties, as usual. The damage by late blight was more severe than usual[416] because of the growth retardation during the early summer. In E Germany leaf roll[417] was rare but 'black leg', in a generic sense, and *'Gipfelrollen'* were common, varying according to the provenance of the seed potatoes (Schander, 1918: 157).

Schander (1917) went into the effects of war conditions, the lack of draught power, labour, and manure. In 1916 the large estates in the east managed somehow to plant at the right time and in the right way. The small farmers there often ploughed when the soil was wet and planted before the soil had dried sufficiently, to the detriment of the potato crop. The modest quality of the seed potatoes used for the 1916 crop was thought to have only a limited effect, at least in the eastern part of the country, though wide planting distances, halved seed potatoes and too small seed potatoes might have reduced yields. Schander, however, concluded that 'natural causes' explained the harvest failure. These reduced early potato growth and stimulated disease development, especially of black leg and late blight. 'Degeneration' was not mentioned in the final conclusions. In the end, the author gave little weight to war-induced limitations of potato cultivation, at least in the eastern part of Germany[418].

Cereals. Wheat diseases were common because of the low temperatures in early summer[419]. A yellow stripe rust epidemic hit the wheat, as in 1914, but its quantitative effects are not clear[420]. In Prussia (present Poland) the rust was sometimes more local than general, as in Bohemia (present Czeckia). In the district Halle yellow rust was unusually heavy, and losses varied from 5 to 20 per cent with extremes up to 50 per cent. Curiously, yellow rust was severe on rye in Moravia (in present Czeckia) and N Lower Austria, and noteworthy on rye in West Prussia (Chapter 2). This yellow rust epidemic on rye was probably the last ever on rye. In Mecklenburg vole damage in cereals was very severe and at places the crop was lost (Zimmermann, 1916b).

5.4.4. 1917 – the frost year

Weather. Severe and prolonged frosts in early 1917 were disastrous in Germany, as in the Netherlands, for a variety of reasons. The waterways were frozen for a long time so that transportation and distribution of coal for heating and of food were hampered for months, to the distress of the civilian population[421]. Railways were overburdened.

Many fall-sown crops were winter-killed. Spring tillage had to be done hurriedly as the spring arrived late. On heavy soils severe crust formation impeded the usual tillage operations (Schander, 1918: 205). The retarded spring ended in an intensive drought lasting well into July. The spring crops that had been planted in time showed a good early growth[422]. The summer of 1917 was clear and dry, and crops were harvested late. Though summer weather was favourable many cereal crops suffered from drought[423].

Potato. The long winter and very dry spring caused delays in planting, even up to 20 June, and in crop emergence. Strong crust formation impeded ridging. Late blight appeared late, mid September 1917, this year on the late varieties only. In the Berlin area, however, most of the potato plants were diseased[424]. Due to the late growing season the tubers had insufficient time to ripen. The tubers remained loose-skinned leading to storage problems. Leaf roll appeared in pure form, but not so frequently, and few other diseases occurred. These notes refer to NE Germany primarily[425]. For Germany as a whole potato yields were relatively good[426].

Cereals. Winter damage in wheat and rye was considerable, as in Bavaria and E Prussia[427]. Due to drought up to mid-July the grain harvest was low[428], especially in Prussia. Little evidence of loss was found except for the gross figure in Table 5.1, indicating that the productivity of cereals in tonnes per ha reached its war-time low in 1917.

5.4.5. 1918 – the meagre year

Weather. The 1918 weather was about normal, with a mild winter and rather cool and dry spring and summer.

Potato. The 1918 potato yields were nearly as miserable as those of 1916 (Table 5.1). I did not find a specific explanation. I tend to believe that the well-known war effects, lack of equipment, draught power, labour and manure, to which might be added a waning human energy with flagging interest in crop care, explain the poor harvest results.

Cereals. In the fall of 1917 voles in rye became a problem in just emerged fall-sown rye in Thuringia[429]. Only in the large fields some crop was left in the middle the field[430]. The fall-sown rye had become very frost-sensitive due to *Fusarium* infection of seed and seedlings during the cool and wet summer of 1917 and the late 1917 harvest[431]. It is not clear whether the damaged 1918 crops were thin only, or lost completely.

S Germany suffered a serious outbreak of voles that caused great damage in spring and early summer, mainly in Saxony and Thuringia but also in Baden, Bavaria, and Württemberg (Schwarz, 1919). In Moravia (present Czeckia) great damage to cereals was done over 18,000 ha by the corn ground beetle[432].

Comment. The cultivation areas of wheat, rye and potatoes had been strongly reduced. Yields of wheat and rye were rather low and of potato very low (Table 5.1), probably due to lack of fertilisers and inadequate crop care.

5.5. Food security in Germany

The Allies were successful with their blockade of Germany at sea. The political pressure exerted by Great Britain on the neutral countries, among which the Netherlands, became so strong that they gradually stopped exporting to Germany[433] (Chapter 4). Within Germany supply to the army had, of course, priority but toward the end of the war even the soldiers went hungry.

The German authorities issued several orders to regulate food production and distribution. The results were disastrous. Delivery problems were frequent leading to long queues in front of the shops. The black markets flourished and became a 'second economy'[434]. Lack of draught power and labour induced farmers to grow less labour-intensive crops that did not contribute much to food production. When fertilisers and manure were not available they often preferred fallowing their land; the productive area was reduced by ~16 per cent (4 million hectares) (Flemming, 1978: 84; Table 5.1).

Consumers woke up after the first sudden increase of food prices, December, 1914. The food situation deteriorated continuously (Flemming, 1978: 78). No wonder that strong animosity arose between townspeople and farmers[435], a feeling that became increasingly confounded with a deepening political fissure between 'left' in the cities and 'right' on the countryside. In the Netherlands, in contrast, the city dwellers' appreciation of farmers increased during the war (§4.5.4). The government produced over-regulation, bureaucracy, inefficiency, ruled against the farmers' interests, and led people into a dead alley (Hirschfeld, 2003: 237). The 'fully incomprehensible economic and pricing policy' was blamed[436]. A solid German agricultural economist described the agricultural policy as 'complete brainlessness'[437].

German food policy was catastrophic. Even the nationalistic Honkamp complained in 1918 about 'inhibitive official measures'[438]. The first public demonstration, with food rather than peace as the issue, was in 1915[439], followed by strikes. Imports of food commodities, e.g. potatoes from the Netherlands and grain from Romania[440], had little impact. Poor weather enhanced the result of an evident neglect of food production by the Central Powers, and the blockade did the rest.

The winter 1916-17 was a hunger winter known as the 'swedes winter' (G: *Steckrübenwinter*)[441]. The scarcity was primarily due to poor harvests, but failure of the price regulation system induced farmers to feed potatoes to their animals or sell it to distilleries rather than to the authorities (Mommsen, 1995: 686) or to the craving population. Excess deaths due to hunger were considerable. An official communication[442] entitled 'Hunger in Germany due to the British sea blockade' read:

> 'According to calculations only 130 grams of protein and 1344 calories were available per person per day in de big cities in 1916, whereas 2569 are desirable, that is about half. In the spring of 1917 the quantities were reduced to 30 grams and 1100 calories. The increase of deaths among the German population amounted to 9.5% in 1915, 14% in 1916, 32% in 1917, and 37 % in 1918. Among children from 6 to 15 years old the increase of deaths was 55%. Accordingly, the Home Department calculated the death sacrifices of the blockade at 260,000 in 1917, 294,000 in 1918 and 763,000 in total over 1915 through 1918.'

The people suffered heavily and the public support for the war, at the onset very positive, decreased accordingly. The awkward plight of civilians was aptly sketched in the bestseller novel by Remarque[443]. The fighting spirit of the soldiers diminished at the thought of the suffering of the folks at home. Aereboe stated[444] 'Correct is only, that the difficulties in feeding the people at the front and in the country after all have broken the stamina of the German people'.

In most cases deaths by famine and by infectious disease accompanying famine cannot be distinguished (Dyson & Ó Gráda, 2002). Fortunately, medical care in Germany was so good that, though infectious diseases popped up occasionally, epidemics could be prevented. Under-nourishment favoured morbidity by pulmonary tuberculosis[445] and mortality by influenza, especially in 1918. The death rate of children (not infants) increased by 50 per cent, over-all morbidity of children by the usual child diseases decreased but child morbidity due to tuberculosis and rachitis increased in the cities[446].

The food situation in Austrian-Hungarian double monarchy and other 'Central Powers' was not better than in Germany (Bruckmüller, 2001). The lack of equity in the access to food led to unrest, protest actions, and strikes[447]. Excess deaths of civilians amounted to 400,000 (Audoin-Rouzeau, 2003). Similar figures (300,000) were quoted for Bulgaria, Romania, and Serbia. Generally speaking, the various countries composing the Central Powers underwent the same fate as Germany suffering want and hunger, disorder and disorganisation. The Central Powers shared military affairs, not food (Siney, 1957: 257). Famines occurred in Galicia (today sprawled over the border Poland-Ukraine) and N Italy, areas then under Austrian rule (Hirschfeld *et al.*, 2003: 565).

In 1917 the USA entered the war and changed the balance of power among the warring parties. In Germany and its associated countries political unrest led to changes in governments. Starvation and hunger undermined the morale of the civilians and, hence, of the soldiers[448]. Germany's military supremacy broke down in 1917 but the final armistice was delayed until November 11th in 1918. The Peace Treaty was signed on 28 June 1919.

5.6. Conclusions

Limiting[449] myself to the German *'Kaiserreich'*, to three starch crops (rye, wheat, potato), and to plant protection problems mainly, I arrive at the following points.
1. World War I had a strong impact on plant protection affairs in Germany, due to lack of labour, crop care, draught power, planting material, fertilisers, and pesticides. War does indeed affect plant disease epidemics.
2. Plant protection problems did affect the course of World War I, contributing to food scarcity, causing malnutrition, hunger and famine, and thus reducing the fighting spirit of the soldiers and the endurance of the German people. Plant disease and pest epidemics do affect warfare.
3. Potato late blight conducted its own *'Blitzkrieg'* in 1916, reduced yields, spoiled storages, and thus added to the starvation of the German people, together with other potato diseases (bacterial blackleg is a candidate) and with pests and diseases of cereals.
4. Late blight and 'degeneration' of potato, epidemic diseases of cereals, outbreaks of voles, and possibly other pests and diseases should, at least in part, be seen as symptoms of one societal syndrome, the gradual but inevitable collapse of a well-organised agricultural production system, due as much to incredible mismanagement as to war scarcities.
5. As to the weather, Germany had bad luck. Honcamp[450] complained rightly '... that during all of the war ... the weather god has not chosen sides with us ...'.
6. History has few one-to-one relationships. During World War I plant disease was important, impressive even, but not in itself decisive. It was but one of the many factors affecting the outcome of the war.
7. Poor governance deepened an inevitable food crisis that killed at least 700,000 people in Germany and a similar number in the other Central Powers.

6. Impact of plant disease epidemics on society, under-rated or over-estimated?

The foregoing chapters taught that a one-to-one relationship between a plant disease epidemic and a social event of historical significance is rare. The impact of plant disease epidemics on society may be underrated or overestimated. Two examples of each case are given. Underestimation is exemplified by the epidemics of ergot on rye, in France, 1789, and Russia, 1722. The first case might have contributed to the French Revolution of 1789. The second case led to an historical non-event. Examples of overestimation are the association of black stem rust on wheat with the Russian Famine, 1932/3, and of brown spot on rice with the Bengal Famine, 1943.

'Les souverains peuvent avoir plus ou moins de puissance, mais ils ont partout les mêmes devoirs.'
The monarchs can have more or less power, but they have the same obligations everywhere.
Malesherbes to King Louis XV, Remonstrance[451] of 18 February 1771 (Badinter, 1978).

'According to the unwritten compact between king and people, in return for their submission, the king promised to assure them their subsistence.'
S.L. Kaplan (1982: 66) on 18th century France.

'Au roi de bonté et de charité, les paysans opposent tous ceux dont ils supportent la domination et les prédations, et d'abord les seigneurs «sans cesse occupés de sucer leur sang»'.
To the king of goodness and charity, the farmers oppose all those of whom they undergo the domination and the predations, and first the landowners 'continuously engaged in sucking their blood'.
Furet & Ozouf (1988: 106) on the preparations for the French elections of 1789.

6.1. Plant disease epidemics and societal disruption

A plant disease epidemic, could it lead to a devastating famine, the fury of a revolution, a lost war, the implosion of a governmental system, the collapse of an organised society? Or should we think of a plant disease epidemic as a symptom only of a society already disintegrating and sliding downhill?

Both views are attractive in their simplicity. In the period covered by the Chapters1-5 little could be done to stop an epidemic once it was on its way. Where people depended on a single crop hunger and starvation were inevitable. Such dependence was in itself a symptom of societal distortion, as in Ireland and in the Bommelerwaard (in the Netherlands), 1845/6. The occurrence of epidemics on two major food crops, rye (Chapter 2) and potato (Chapter 3) in one year – 1846 – was really a stroke of bad luck. A well-organised society might attenuate the ill consequences of an epidemic by opening food stores, importing grain, regulating food prices, forbidding hoarding and speculation, and so on. All this was done in 1845/6, sometimes too late and always too little. The rural masses could hardly be reached. Several governments, unable to cope with the emergency, imploded. The epidemics were among the proximate causes of the implosion but the ultimate causes had little to do with agriculture (Chapter 3).

During World War I the weather gods took side against Germany but neither did they spare the Netherlands. The German government made a mess of food security (Chapter 5); the Dutch government muddled through and did not do so badly (Chapter 4). The neutral Netherlands had the foresight to store grain (e.g. rice) in advance but belligerent Germany expected to take food from the areas conquered in a brief war. In both countries plant diseases made their impact on major starch crops, in Germany more than in the Netherlands, but there is little evidence of decisive epidemics.

It seems as if plant pathologists experience difficulties to value the societal effects of plant disease epidemics, sometimes jumping at conclusions and either under-rating the effects of 'their' epidemics or over-estimating them. The present chapter provides two examples of each case in a tentative manner as the necessary documentation is not (yet?) available.

6.2. Plant disease and the Great French Revolution, 1789

6.2.1. Unrest in 18th century France

For rural France the 1780s were a miserable period with poor weather, meagre harvests, and wide-spread social unrest. Taxes weighed heavily upon the French peasants, among which land rent in money or in kind, the King's rights, the Church's tithes, import/export duties at provincial borders, levies to enter town or market place, tolls at the many turnpikes, milling rights, and the hated *'Gabelle'*, the tax on salt. Needless to say that the nobility and the Church, both rich in land, enjoyed tax exemption.

Peasants were not free to move. Many were, in fact, serfs. They were obliged to work a certain number of days per year for the landowner. The latter enjoyed hunting rights. The hunters could chase through the crop, even when ripe for harvesting, without compensating the farmers. The humiliations were difficult to bear. Discontent was general, disturbances were frequent. In years when food was really scarce just before harvest people became panicky, rumours flew over the countryside, and a curious kind of 'fear' (F: *peur*) arose, a panic (as also in 1845/6, Chapter 3).

The harvest of 1788 had failed in France, and had been poor in most of W Europe. The winter 1788-89 had been unusually severe (Figure 6.1; Le Roy Ladurie, 1967: 621). Food scarcity reigned and people were discontented. The increase of the food prices and the decrease of job opportunity for labourers and artisans led to unrest and strikes, as e.g. in Germany (Dirlmeijer, 1995: 232). In the Netherlands bread prices had to be regulated and even subsidised in 1789 (Anonymous, 1789).

In France the situation was worse. The rural poor lived under horrible conditions (Goubert, 1974). The French had to import grain from the Baltic area, by way of Amsterdam, for 'the in high revolution fever burning Paris' (Brugmans, 1937: 28). This transit trade was much against the wish of the Dutch populace that was in serious need of grain too, but the weakening of France was politically welcome to the Dutch Republic.

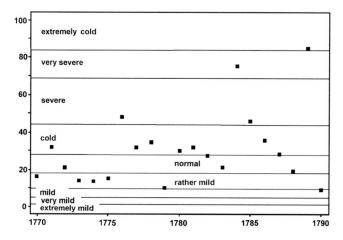

Figure 6.1. The Frost Index for the Netherlands in years 1770-1790 (IJnsen, 1981).
The value for e.g. 1780 should be read as the value for the winter 1799/80. The winter 1788/89 was extremely cold, the winter 1789/90 was mild. Horizontal – years. Vertical – Frost Index (ranging from 0 to 100).

In several districts of France the wheat had been extremely affected by bunt during the years 1784-87, to the degree that threshers fell ill, even died, due to the dust of the bunt spores (Rozier, 1809: 488). The 1788 harvest had been so bad that food shortages appeared on the country-side in September already (Lefèbvre, 1932). Where the townsfolk may have had enough money to buy food, scarcity reigned in the rural areas. After the severe winter 1788-89 people must have been desperate. Bands of errant beggars appeared, the 'poor of the night'. Standing crops had to be guarded, lest the corn be stolen before ripening, 'cut green' (F: *coupé en vert*). In the towns the poor rioted in the spring of 1789. The grain harvest of 1789 was bad again, two thirds of normal or even less[452].

6.2.2. The Great Fear

Fear spread over the country. Farmers feared beggars and robbers. Armoured bands roamed about. Townspeople feared country people, *vice versa*. Paris feared the provinces, nobles feared the king. Farmers hated the nobles, who lacked cash and tried to squeeze money out of them. Nobles feared the farmers, who refused to pay the inordinately high taxes. In the end, the 'people scared itself' (F: *le peuple se faisait peur à lui-même*), in a way hardly imaginable today. The army weakened and anarchy grew. Towns, villages and individuals took to arms, the alarm clock (F: *tocsin*) was in readiness. Autosuggestion played a role in nightly panics that built up gradually into local 'fears' as of May, 1789. A xenophobia arose and rumours about foreign troops began to circulate. The aristocracy was thought to plot with foreign powers. The nobles were suspected to hire errant men. Aristocrats mistrusted their servants and several fled to neighbour countries. Meanwhile the bread became so expensive that new insurgencies burst out in July, 1789, the 'grain troubles' (F: *troubles frumentaires*).

The situation was highly inflammable. King Louis XVI was in fear too. The state had run out of money and it was not feasible to raise taxes once more. In his anxiety the king had summoned the States General, to meet on 15 May 1798, for the first time since 1614. They consisted of three States (F: *États*) or chambers, nobility, church, and bourgeoisie. The peasants and other country folks, the Fourth State so to say, were not represented, of course not.

While the gentlemen at Versailles, near Paris, were discussing the reorganisation of the French state the panic on the countryside reached its boiling point when everybody became afraid of everybody else. People took to arms, guns, scythes, pitch-forks, or sticks with stones tied on. A lonely wanderer in the twilight could arouse a village alert. Women and children left their homes to hide in the forest, taking along some food and a few valuables. A group of cattle in the dark was seen as a band of dreaded 'brigands', ready to attack the hamlet. An ordinary fear became the 'Great Fear' (F: *la Grande Peur*) of the French historians (Lefèbvre, 1932).

There was, also, serious rioting. A few castles were set in flames. Several tax archives were destroyed by infuriated peasants. The gentlemen assembling in Versailles, nobles, clergymen, and bourgeois, panicked in turn. Progress had been slow but in face of the 'danger' the congregants experienced a shake-up.

On 14 July 1789 the Paris rabble had stormed the Bastille, the hated prison-castle downtown (presently, 14 July is National Day of France). The king, deeply impressed, became more positively inclined toward some of the revolutionary ideas. Information about the rural panic and the various disturbances in the country was, of course, received in Versailles. In response, the *Constituante* (Constituent Meeting) decided in the night of 4 to 5 August to abolish the feudal system, including serfdom, and privileges of clergy and nobility. This was a crucial step in a lengthy process called 'The French revolution'.

In times of want such fears, fed by a 'famine plot persuasion' (Kaplan, 1982), were not uncommon in France. But when does a 'common fear' pass into a 'Great Fear'? When the state, the king, the nobility ran out of cash and could no longer press money out of murmuring if not mutinying peasants? When the government, loosing its grip on state affairs, collapsed? When everybody turned against everybody, often with weapons at hand? Suddenly, beginning ~20 July, a panic spread over most of France, sometimes as fast as a horse could gallop. The Great Fear had a few focal points from where it spread in various directions[453]. What happened?

Comes in Matossian (1989), the historian who studied food poisoning by naturally occurring fungi. She reasoned that ergotism made the French rural population so suggestible that an 'ordinary 'fear' changed into the 'Great Fear'. Here I recall some of her points. (1) The weather in the early summer of 1789 had been very favourable for ergot on rye. (2) The Great Fear occurred only where rye was the main crop. The areas with maize, potato, or buckwheat as main crops remained unaffected. (3) Miscarriage and mental illness were unusually frequent during the summer of 1789. (4) The spread of the panic showed a pattern, with a number of foci that is difficult to explain. Its N to S movement may be associated with the pattern of the rye trade, infested rye from N France being carted southward (*ibid.*: 84). Curiously, Barger (1931), the ergot specialist, did not mention 1789 on his list of typical ergot years. Bouchard (1972: 101), chronicler of ergot stricken Sologne in MW France, mentioned 1789 only as a year with poor grain harvests.

Matossian (1989: 82ff) combed the literature for ergotism in 1789. Her harvest was meagre but positive[454]. There was much ergot in N France and in parts of England, and a case of ergotism in N Italy. In some areas of France miscarriage was frequent during the summer months, a more pertinent sign of ergotism. One physician 'chronicled a marked deterioration in public health in the second half of July' that he attributed to the consumption of 'bad flour'. In Paris a clear increase of nervous diseases in the second half of July was attributed, again, to 'bad bread'.

The timing of the Great Fear relative to the grain harvest is of interest. It was harvest time, a capital date in rural life (F: *'date capitale de la vie rurale'*; Furet & Richet, 1973: 86). Unfortunately, few precise data were found. In the Dauphiné the stewards of the estates considered the Great Fear an undesirable interruption of the work [harvesting?] on the land (Palou, 1955: 53). In the Poitiers the Great Fear interrupted fieldwork at harvest time, 28 July, in one place; in another place the panic began at the peak of the grain harvest (F: *en pleine moisson*; Diné, 1951: 118, 122). These tiny bits of information are not incompatible with the idea that unripe, ergoted rye had been eaten just before the outbreak of the Great Fear.

Matossian (1989: 87) summarised her findings: 'it appears likely that certain ergot alkaloids created a suggestive state of mind among the rye-eating populations'. Cultural factors would then shape the associated visions such as 'brigands coming to steal crops'. She concludes 'The Great Fear of 1789 was among the last of the bizarre events in European history that may be explained by food poisoning'[455]. Ergotism may have been instrumental in the shift from an 'ordinary fear' to the 'Great Fear'. Thus it may have lead to the famous 'oath at the fives-court', sworn by the parliamentarians at Versailles in the night of 4 to 5 August, to abolish the *'Ancien Régime'*.

6.2.3. Studies on ergotism

The history of ergotism, reaching back for millennia, has been rendered by among others Barger (1931) and Chaumartin (1946). The European population was struck by terrible epidemics time and again. The poor suffered most. The two major forms are gangrenous (Box 6.1) and convulsive ergotism. France has been plagued by ergotism from the Middle Ages onwards. Intuitive knowledge linking symptoms to ergot existed, without scientific proof. A more solid link between the symptoms of ergotism and the ergot of rye was laid around 1670 by a French country physician named Thuiller (Carefoot & Sprott, 1967: 24), but his ideas had not been accepted by everybody. Present knowledge on ergot is summarised in Boxes 6.2, 6.3 and 6.4.

Box 6.1. Symptoms of gangrenous ergotism as described by Tessier in 1776 (Bouchard, 1972: 112/3; own translation).

'The sick persons, especially those better off, experienced headaches and stomach pains during the first two to three days. The fever came, they all felt painful fatigues in the lower extremities. These parts swelled up without apparent inflammation; they became rigid, cold and pale, and gangrened. Sometimes they oozed a stinking cruelty, or drops of blackish blood; sometimes worms were formed [?]. Ordinarily, a narrow band of light inflammation was above the gangrene, where it was bounded, and where later on the limb was separated by itself. The gangrene began at the centre of the part, and only appeared on the skin a long time after. The toes fell off first and successively all the joints detached. The upper extremities, though more rarely, underwent the same fate. Unlucky persons were seen whom nothing remained but the trunk, and who survived another few days in that state. The limbs detached by themselves without bleeding. It happened sometimes, that in stead of detaching, they emaciated, and detached without falling apart in dust. The persons affected by that disease were stupefied; they had a swollen belly, a light and rapid pulse.'

Box 6.2. The life-cycle of ergot, caused by the fungus Claviceps purpurea Tul. Data from Butler & Jones (1955).

The life cycle of ergot (Figure 6.2) has been elucidated primarily by the Tulasne brothers (1853: 45). Ergot is caused by a fungus that infects the inflorescences of various grasses and cereals. On rye it appears in its most conspicuous stage as *sclerotia*, black kernels ~10-25 mm long and 5-8 mm wide, replacing rye kernels in the ears. In rye, usually 1-2 sclerotia per ear are found. These sclerotia ripen at the same time as the kernels. The ripe sclerotia fall to the ground or they are 'harvested' with the grain. If no precautions are taken they are milled with the grain or sown with the rye seed in early fall.

Sclerotia on or in the ground easily overwinter, resisting severe cold. In the spring, when favourable conditions prevail (Box 6.3), sclerotia on or just under the soil surface germinate. They produce ~6 club-shaped stalks of 5-25 mm long (called *clavae*) with a pink cap (a *stroma*) in which *perithecia* are formed that produce *ascospores* after ~6 days.

The ascospores are extruded as a viscid mass allowing dispersal by rain splash or insects, or alternatively they are produced dry and forcibly ejected up to 20 mm high to be dispersed by turbulent air. Ascospore production is somehow synchronised with early anthesis. Open rye florets remain infectable up to ~14 days. Ears just flowering are the most vulnerable.

The ascospores land and germinate on the young stigmata of the rye florets, the primary infection. A *germ tube* grows toward the ovarium where it forms a *mycelium* that produces 'honeydew', a sweetish, sticky, yellow fluid loaded with *conidiospores*. The honeydew attracts insects that disperse the spores from floret to floret and from ear to ear, thus causing secondary infection. The spores can be splash-dispersed during rain showers. They can also be mechanically transmitted from ear to ear by direct contact. The conidiospores germinate and infect the ovarium that is converted into a sclerotium (called *ergot*) colouring from white over orange-red to purplish black.

Field edges bordering verges with infected grasses and low lying humid parts of the rye field are the most vulnerable parts of the field. Irregular crops, sown too thin or damaged by frost, are particularly at risk because infectable young ears appear over a prolonged period.

The fungus has various races of which some attack rye. Among the rye-infecting races exist strains with different alkaloid content, varying according to geographic area and historical period. Some strains cause gangrene in humans, others cause convulsive ergotism. Rarely a strain causes both types of ergotism.

Figure 6.2. Ergot on rye – Claviceps purpurea (Fr.) Tul. From J.B. F. Bulliard (1791), Herbier de la France. Seconde division. Histoire des champignons de la France. II. Paris, Didot. Plate 111.

A.	*Spikelet.*	*E...F.*	*Various shapes of ergot.*
B.	*Spikelet, one flower removed.*	*M.*	*Flowering spike.*
C.	*Ovary*	*N.*	*Spike [ripe] with good kernels and ergots.*
D.	*Kernel, two sides.*		

Box 6.3. *Environmental conditions favouring ergot infection on rye and ergot control (Data from Butler & Jones, 1955 and others).*

Environmental conditions

Sclerotia	A moist environment in the preceding fall.
	Winter damage creating an open crop with great variation in flowering time of main, secondary, and late tillers.
Germination	Moist spring with minimum temperature ~11 °C and maximum temperature of ~18 °C.
Ascospores	Extrusion requires high humidity, RH >77%.
	Dispersal by insects, rain splash, and by direct ejection and turbulent air stream.
Conidia	Production during cool and moist weather, RH ≥74%, Temp ≤13-15 °C.
	Dispersal by insects, rain splash, or direct contact.

Control

Use healthy seed	Eliminate sclerotia by hand-picking or brining.
	With hand-picking fragments of broken sclerotia may be missed.
	With brining in salt water floating sclerotia are removed, sunken rye kernels are washed and dried.
	Use old seed; sclerotia age and become ineffective within 2-3 years.
Avoid infection	Early mowing of field borders, especially in wet, sheltered places, removes ergot on grasses that can infect rye.
	Rotation, 2-3 years without rye eliminates infective sclerotia left in the soil.
	Deep-ploughing buries infective sclerotia.
Crop husbandry	Produce a rather dense, regular crop.
	A dense crop hampers dispersal of ascospores, but may further dispersal of conidia.
	A regular crop with all ears heading and flowering at the same time reduces the infectable period of the rye.
	Don't use growth regulators.

The French Royal Society of Medicine, wanting to bring the confusion around ergot and ergotism to an end, requested father Tessier, professor of Medicine at the Sorbonne, Paris, to study the problem and write a report. Tessier went to Sologne in July 1777. He reported the results of his studies e.g. in his treatise on cereal diseases (Tessier, 1783). Sologne is the district south of Orléans, between the rivers Loire and Cher. It is a sandy area, generally humid, with large forests and many ponds. The inhabitants, the Solognese, were poor and always had too little to eat, though their diet was varied. They ripened late and died early, suffering annually from intestinal diseases and malaria (Bouchard, 1972).

Sologne was known as the area with endemic ergotism. 'The inhabitants of Sologne, where the gangrenous disease occurred most frequently, and where the ergot is more abundant than elsewhere, live, during the first three months following the harvest, only on bread made from rye, the bran included'[456]. The Solognese knew the relationship between ergot and disease (Bouchard, 1972: 105) but did little about it, out of indifference but also for a technical reason. Ergots in rye, if not lost during harvesting, could be hand-picked because of their

---

Box 6.4. The effects of ergot poisoning in humans. Data collated from Matossian (1989), Van Genderen et al. (1996), and Wikipedia.

Ergots contain 0.02-1 per cent alkaloids of the indol-alkaloid type, the ergolines. Ergolines usually have two parts, a lysergic acid amide and an amino acid. Lysergic acid is the base material for the production of lysergid or LSD. Some 30 different ergoline molecules are known.
A dose of 5-10 g of fresh ergot kills an adult person. The no-toxic-effect-level is 0.1 mg per kg body weight. The no-intervention-level is 1 ppm.
Ergotamine leads to vasoconstriction. Ergometrine leads to contraction of the uterus. Hence, ergot was used in medicine to stem bleeding after child birth. By the same token it can be used as a birth inducer and as an abortive. Ergot has also been used to correct hypotonia at morning rising.
Ergot enters the food chain when the sclerotia have not been eliminated before milling the rye. This happens by, sometimes criminal, negligence or by the hurry to get food after a summer period of scarcity.
Symptoms of ergot poisoning vary considerably. Light symptoms are indicated as itching of the skin (G: *Kriebelkrankheit*). Acute poisoning affects the central nervous system, with vomiting, headache, confusion, hallucinations, fear, convulsion, and unconsciousnous (convulsive ergotism).
Chronic poisoning leads to vasoconstriction, ischaemia (cold skin), heavy pain in hands and feet (Saint Anthony's fire). Severe poisoning leads to gangrene (gangrenous ergotism, Box 6.1).

---

colour (purplish black) and size (up to one inch). The Solognese version of ergot, however, used to be of the same size as the rye kernels (Tessier, 1783: 25) so that mechanical removal was hardly feasible. The alternative, brining the grain in salt water and floating the ergots (the rye kernels would sink) was out of reach as salt was heavily taxed and expensive. Sologne was a well-chosen study area.

6.2.4. Tessier's studies

Many grasses were ergoted but among the cereals it was mainly rye (Bulliard, 1783: 49). Linnaeus (1751: 243) classified ergot (L: *clavus*) as a plant disease without further comment. Fabricius (1774: 48) ranked ergot under Class V 'Injury', Genus II 'Galls', thinking it the result of an insect puncture, a current idea at the time. Darwin (1800: 322) correctly thought ergot to be an 'internal disease'[457]. The educated people of the 18th century knew about the relationship between ergot and the 'itching disease' (G: *Kriebelkrankheit*), some with a certain degree of disbelief[458].

If the 17th century was the age of classification the 18th century was the age of experimentation. H.-A. Tessier (1741-1837), physician and agronomist, reviewed the already extensive literature on ergotism (Tessier, 1783). He was dissatisfied with the quality of the experiments recorded so far. No starting conditions were specified, no controls were available, and no witnesses were called in. The results were contradictory. Tessier began a series of agronomical and toxicological experiments.

The appearance of ergot on rye varied somewhat per region. The rye in Sologne usually had rather small ergots, four to five per ear, sometimes ten to twelve, rarely twenty. The secondary tillers of winter rye, smaller and damper, had more ergot that primary tillers. Tessier[459] knew that the amount of ergot varied according to season and locality, more in humid places and on fields recently reclaimed after fallowing. The year 1777 was wet with much and 1780 dry with little ergot. Frequent ploughing reduced the risk of ergot. Whereas *'méteil'*, the mixture of wheat and rye had relatively less ergot than rye pure, *'hyvernache'*, a mixture of rye and vetch had more, the rye roots kept moist by the vetch.

Tessier had seen the shiny, viscous, sweet sap on the ears, with the taste of honey, even on the glumes that enclosed nascent ergots. He observed the honeydew, but he was not sure about its meaning since he saw ergot also on ears without honeydew. Tessier noticed the white substance replacing the rye kernel that became a cockspur-like, violet ergot eight days later. He measured the growth rate of ergots as ~2 mm per day.

For the causation of ergot Tessier quoted several explanations but he wanted facts, not reasoning. So he performed various experiments. He punctured flowers with a pointed knife, without result; thus he excluded puncturing insects as a cause. He tested various soils and different irrigation regimes. The poorer the soil and the wetter, the more ergot he obtained. In his 1780 experiment (the dry year) the irrigated plot, measuring 1.20x5.12 m^2, produced eight times more ergot than the dry plot; he made no replications. Watering the ears did not promote ergot. He concluded that ergot was due to the wetness of the soil. A [further] explanation of ergot was unknown and unnecessary, he declared.

The results of toxicological experiments published so far were confusing and had to be repeated. Tessier performed several experiments, sometimes repeating those of predecessors, but taking extra precautions such as using animals of different species, healthy, in their prime, in good accommodations, with good care, and he invited witnesses. Dosages were up to one quarter ergot, as with humans. A duck fed with ergot died within a week, a turkey needed three weeks to die, one hog six and another hog 10 weeks. A six weeks old hog died with convulsions after three weeks. A strong hog of six months old did not like his ergoted porridge but ate it. Five days later he had red eyes, on day 13 he was dizzy, on day 69 he died with convulsions. A young dog disliked his ergoted porridge too much and did not die[460].

Whereas the experimental animals were healthy at the start the Solognese people were ill-fed and weak (Bouchard, 1972: 108), and breathed noxious vapours[461]. The disease was not contagious, affecting men more than women. With ergotism pregnant women aborted and nursing women dried up. The disease killed; of the 120 patients in the hospital of the town Orléans only four to five recovered[462].

Tessier was well aware of the medical applications of ergot, though somewhat sceptical, inducing or precipitating childbirth, staunching the flow of blood. The use of ergot as an abortive was a well-known secret. The Reverend Berkeley (1860: 66) commented primly 'Ergoted grain, however, which ows its origin to a closely allied Fungus [*Cordiceps*], is a most valuable medicine in the hands of the regular practitioner, though often grievously abused from the specific action on the womb'.

The proposition that ergotism changed the existing 'fear' into a 'Great Fear' is attractive. In times of want the rye was harvested prematurely, dried, threshed, and eaten, ergots and all, the bread often inadequately baked. The timing of the Great Fear fits, approximately, with the harvest time of rye in France.

6.2.5. Conclusions

1. The evidence for ergot and ergotism in 1789 is scanty but Matossian's reasoning remains attractive. Except, maybe, in Sologne the symptoms of ergotism appeared to be mild, with mental stress primarily.
2. The frequent observation in 1789 that a local community was affected instantaneously and as a whole points to ergotism, as do the visions, hallucinations, and fears on record.
3. A relatively mild intoxication by ergot in July, around harvest time, may have made the difference between an 'ordinary fear' and the 'Great Fear' of 1789.
4. The Bastille in Paris had been stormed on 14 July and the Great Revolution was already on its way. The Great Fear, and hence ergot, seems to have been instrumental in speeding up the revolutionary process.
5. If so, a plant disease contributed to a sudden and unexpected political phenomenon, the decision of the States in the night of 3 to 4 August 1789 to abolish the privileges, among which tax exemption and serfdom, of church and nobility.

6.3. The war that did not happen, 1722

Peter the Great (1672-1725), Czar of all Russians, was a visionary man. He wanted to modernise the then quite backward Russia, opening his self-contained country to the West. He needed access to seaports for trade. To this purpose he successfully fought the Swedes and in 1703 he founded St Petersburg, on the river Newa, looking West over the Baltic Sea. At the time Russia had no free access to the Mediterranean or the Black Sea. The Ottoman Empire stood in its way.

In 1722 Czar Peter assembled a large army near Astrakhan, where the Wolga discharged into the Caspian Sea. The plan was to march along the Caucasus and to attack the Ottomans in the rear. Some 25,000 troops were ready to go. The army, men and horses, had to be fed. Rye was carted in from a large area, the fresh harvest of 1722. Then, the unexpected happened. Men and horses fell ill, with terrible convulsions. Death followed soon and many men who survived lost hands, feet, or even limbs. The disease was not contagious but the losses were so severe that the campaign had to be cancelled.

Czar Peter had the case investigated by the German physician Schober, who published a note in 1723[463]. The ambassador of France at Moscow, Mr Campredor, mentioned the case to the French ruler King Louis XIV in a letter dated 29 January 1723 (Barger, 1931: 80/1; Box 6.5).

Two independent witnesses gave a description that leads to one conclusion only, ergotism! It was a case of acute poisoning by freshly harvested ergot. At harvest time the toxin concentration is at its highest. Apparently, the 1722 poisoning caused the gangrenous and convulsive forms simultaneously, as occurred a few times more in Russia (Barger, 1931: 81)[464]. Ergot did the army in. The Ottoman Empire was saved. Had the Czar been successful, the political maps of Europe and the Near East would have been very different from their present state.

The Russian side of the story has a different ring[465]. The 'Persian campaign' of 1722-23 under the command of Peter I had the objective 'to help the peoples of Transcaucasia to liberate themselves from Iranian domination, establish Russia's trade relations with the Orient, and prevent Turkish aggression in Transcaucasia.' In July 1722 the Russian army, with 22,000

Box 6.5. From a letter by embassador Campredor to Louis XIV, King of France (from Barger, 1931: 80/1; own translation).

'I believe that the Czar is too prudent to engage in a war that would considerably reduce his forces, whatever success that war could have. All the cavalry, he had brought to Astrakan, is ruined, and his finances are in a very poor state. The bad harvest of the past year [1722], the prodigal quantity of grain perished on the Caspian Sea, will make the provision of the stores difficult, and over twenty thousand persons died already around Nijny [Novgorod?] due to the scarcity. First it was thought that it was the plague, but the physicians who had been sent, have reported after a very precise examination that this disease was not contagious, that she came forth only from the bad grain, that the men have eaten. It was reddish and rather resembles the [intoxicating] ryegrass [*Lolium temulentum* L.], having been spoilt, as is thought by poisonous fogs. The persons, the moment they have eaten from this bread, had become dizzy, with great contractions of nerves, so that those, who did not die that day, have lost their hands and feet, that fell off, as happens in this country, when those limbs are frozen. None of the remedies, applied with infectious diseases, did work on the patients, and only those who had taken good food and eaten different bread, have escaped. The dissertation which the doctors have made at this occasion, is quite interesting, and if I can get a copy, I will have the honour to send it to Your Eminence. Well, as this disease can have odious consequences, by the difficulty to find good rye for the subsistence of the inhabitants and of an army and by the quantity of the bad [corn], ordered to be burnt and that for the rest the events of a war against the Turks could weaken suddenly and without resources the forces and the consideration of the Czar, it is evident, at least up to the present, that the movements, that he did make his troops, have as the prime motive only to show off to the Turkish envoy who is induced to proceed very slowly.'

infantrymen, joined by about 22,000 cavalrymen, moved into S Dagestan. 'In the fall, heavy storms disrupted the supply of the Russian troops, and Peter I was forced to abandon the campaign on Baku and returned to Astrakhan ...' 'The success of the Russian troops and the Turkish invasion of Transcaucasia in the spring of 1723 forced the Iranian government on September 12 to conclude the Treaty of St. Petersburg of 1723', giving up a wide territory among which Baku.

An event of historical dimension, a clash between Russian Orthodox and Ottoman Muslim armies, did not happen[466]. The rest is silence.

6.4. Black rust of wheat and the Russian famine of 1932-1933

6.4.1. The epidemic of 1932

In 1932, a severe epidemic of black stem rust (*Puccinia graminis* f.sp. *tritici*) devastated the wheat crops in many East European countries. This epidemic is well documented (Hogg *et al.*, 1969; Scheibe, 1933; Zadoks, 1965). It began in Bulgaria, strongly affected Romania (Savulescu, 1953: 144), spread westward through the Danube valley and eastward to the Ukraine, rounded the Carpathian mountains and touched Poland, Germany and some Scandinavian countries. An easy guess relates this epidemic to the Russian famine which peaked 1932-1933, with an excess mortality of millions of people.

This extremely severe 1932-33 famine in the Soviet Union was studied in detail by Russia specialists Davies & Wheatcroft (2004), who had access to the Russian archives. They did not ignore the stem rust epidemic but gave it a footnote only. Far more important, in their view, were the dekulakisation (R: *kulak* = prosperous farmer[467]) and the collectivisation, imposed upon Russian and Ukrainian agriculture and forced through in a very short time. The communist rulers knew next to nothing about agriculture and they destroyed the existing infra-structure without establishing an adequate new one. Farmers hardly cooperated.

Everything went wrong in 1931 and 1932. The 1931 weather was unfavourable with summer drought and so much rain in the fall that the potatoes could not be lifted. Seed was delivered too late. Hence the wheat sowing was delayed. Several wheat diseases, among which primarily rusts, insects and weeds could damage the crops. Supposedly, the severe black rust epidemic of 1932 struck the Soviet Union heavily. Elements contributing to the severity were retarded delivery of seed in the fall, a late and cool spring, delays in sowing spring wheat due to poor weather, and heavy summer rains. The Central Russian Plateau near Kursk was hard hit, possibly because of its somewhat higher altitude (>200 m), which might have delayed ripening so much that an extra generation of rust could develop in comparison to its lower surroundings.

As a pathologist I was inclined to see the rust as the proximate and major cause of distress but this view is contested.

6.4.2. Industrialisation of Russian agriculture

At the time, the Soviet Union went through a period of spasmodic change. Industrialisation was imposed at a high rate as was collectivisation of agriculture (Davies & Wheatcroft, 2004). Peasants were forced into collectivisation. Some 2 million people were expelled from their lands. Those who resisted were punished, often by deportation. The rate of change was so high that agriculture became completely disorganised. Compulsory area extension conflicted with necessary crop rotation, and led to reduced yields. Draught power failed since horses died by lack of feed, and they were not replaced in time by tractors. The quality of tillage was poor, too late, and/or too superficial. Weed control was miserable. Voroshilov, member of the Politburo, reported on 26 July 1932 that the North Caucasus gave 'a depressing picture of the scandalous infestation of the grain with weeds', an impression confirmed by Western observers. Harvesting was delayed by late ploughing and seeding, by rains, and by lack of machinery or draught power. Finally the weather was not conducive to high grain yields.

Davies & Wheatcroft (2004: 434) clearly stated that the fundamental cause was the 'unremitting state pressure on rural resources'. '… Soviet leadership which was struggling with a famine crisis which had been caused partly by their wrongheaded policies, but which was unexpected and undesirable. …'. 'They [the policies] were formulated by men with little formal education and limited knowledge of agriculture'. 'Above all, they were a consequence of the decision to industrialise this peasant country at breakneck speed'. In short, the political pressures of that particular period, with intentional societal disorganisation, were the ultimate cause of the famine. The then current terminology was 'machinations of class enemies and inefficient organisation'. Biotic actors such as weeds, locusts and, maybe, rust were at best proximate causes enhancing rather than triggering the crop losses.

Yields were good in 1930, the last year before the collectivisation of agriculture. In 1931 the grain yielded poorly, in part due to adverse weather. In 1932 yields were even worse, but in 1933 they recovered. Poor yields and disorganisation caused an incredibly severe rural famine, from spring 1932 until harvest 1933, during which an estimated 5.7 million people died, especially the children, the sick and the aged. Cases of cannibalism occurred by the hundreds. Apart from the incredible suffering and despair of the population[468], recruitment plummeted with long-lasting effects, childbirth going down drastically in 1933. As usual with famines infectious diseases, among which primarily typhus, took their toll.

The worst came in 1933, a rural famine in the main grain areas with mortality rates up to nine times the ordinary rate (Volga area). The famine was too general and too bad to remedy. Russia suffered immensely. According to the historians, it was intentional societal disorganisation that was the ultimate cause of the famine. Plant disease was hardly thought instrumental and it was not considered a proximate cause of the famine.

Today, I see good reason to invert the argument. The disorganisation led to such delays in crop development that the wheat crops became an easy victim of the rust blown in from the SW.

6.4.3. Conclusions

1. In 1932 there was simultaneity of a wheat rust epidemic and the beginning of a famine.
2. The precipitated collectivisation of Russian agriculture caused an incredible disorganisation of agriculture that led to poor crops and low yields.
3. The resulting famine was typically man-made and, according to some, intentional.
4. Delays in planting, crop development, and harvesting allowed the rust to intensify and develop into a severe epidemic.
5. The black rust epidemic probably was an unintended outcome of intentional disorganisation rather than the cause of the general distress.

6.5. Brown spot on rice and the Bengal Famine of 1943

6.5.1. Social order in Bengal

Bengal, presently Bangla Desh and the state of West Bengal in India, but in 1939-1945 a province of British India, was a land of ease where life centred around rice. The social system was relatively simple. Landlords, often dignified as king (B: *rājā*) 'maintained legitimacy primarily by fulfilling the conventional code of conduct deemed appropriate for kings (B: *rājā-dharma*)'. The king maintained a 'court' and was indulgent to his subjects, the rice-producing tenants. The tenants were expected to pay rents to the landlord and to show him deference. The landlord was expected to protect the tenants and their dependents, to be indulgent, and to feed them from his own stock in times of dearth (Greenough, 1982).

In this idealised picture the king was the 'destined provider of subsistence' (B: *annadātā* = rice giver) for 'dependent persons requiring nurture'. Among the dependents were the tenants, often Muslims, but also the Brahmans, healers, astrologers, fortune tellers and caste genealogists. The king, usually a Hindu, might give grants to Muslim scholars and privileged-tenure lands to persons or institutions of great learning.

Not all authors share the rosy view of a relatively happy, decentralised society, only incidentally affected by misfortune. In fact, the social situation had gradually deteriorated under the pressure of imported capitalism. Bailey (1945) observed that peasants were 'living perpetually on the verge of starvation', extremely poor, on holdings too small to make a decent living. One major misfortune was associated with plant disease, the 'Great Bengal Famine' of 1943. Millions of people died of starvation (1943) and of its corollary of infectious diseases, among which cholera, malaria and smallpox. Estimates vary from 1.5 over 3 to 3.5 million excess deaths in the years 1942 to 1946. A birth deficit was most marked in 1944[469].

6.5.2. Brown spot in 1942

The plant pathologist Padmanabhan (1973) was an eye witness of the 1943 starvation and death when reporting for duty at Dacca in October, 1943, at about the peak of the famine. He vividly depicted the cause of the Bengal Famine. The main rice crop, the *aman* crop harvested toward the end of the year, failed in 1942, with losses of 50 per cent and more relative to the 1941 crop. Padmanabhan *et al.* (1948) described the disease in its normal, mild and relatively harmless form and in its aggressive, epidemic form. He gave no explanation of the change from the one form to the other[470].

Padmanabhan (1973) studied the weather data of 1942. From September through November rainfall, cloudiness, and humidity were higher than usual. Apparently the very dark and wet fall of 1942 prompted widespread and severe occurrence of the causal fungus *Helminthosporium oryzae* (*Cochliobolus myabeanus*) on rice in Bengal. This common fungus, causing the disease called 'brown spot', is a relatively weak pathogen which seizes its opportunity when poor soil condition and/or radiation deficiency hamper the growth of the rice (Klomp, 1977). Since rice needs high light intensities to produce a good crop[471] part of the damage might even be ascribed to the radiation deficiency during that particular season[472].

A comparison of yields from variety trials on two locations, with up to 23 varieties, showed that losses during the epidemic season 1942 varied from 10 to 90 per cent relative to the yields of the normal season 1941 (Padmanabhan, 1973). The data show convincingly that the disease was present and damaging. These two trials, supposedly the worst affected ones, were part of a network of trials (Tauger, 2003). Should we believe that the other trials were equally affected? I don't think so.

I would not have questioned Padmanabhan's analysis of the immediate cause of the rice crop failure if he had not mentioned himself that he was not on the spot in 1942. Padmanabhan was a highly competent plant pathologist but his conclusion was based on the analysis of observations from two experimental fields made by others. His detailed studies on losses due to brown spot date from the years following the debacle.

In the treasure trove of rice diseases written by Ou (1985: 202) we read 'The most dramatic aspect of the disease so far recorded was that it was considered to be a major factor contributing to the Bengal famine of 1942 (report of the Famine Inquiry headed by Sir John Woodhead, 1945), the losses then amounting to 50-90% (Ghose *et al.*, 1960; Padmanabhan, 1973).' Unfortunately, I had no access to information on the varieties grown in the farmers' fields, the area covered by each variety, the disease severity per variety, and the geographic extent of the diseased varieties. Nor did I find detailed information on the effects of the radiation deficiency, the storms and the floods. What were the effects due to the lack of manure or fertiliser in times of war (as in Chapters 4 and 5), or of reduced crop care during a period of trouble (as in Chapter 5)? I have a suspicion that the loss figures began their own life foregoing reality. In fact, the 1942 *āman* or winter crop yielded ~83 per cent of the mean of the four preceding years, not enough – of course – but not really bad[473] (Sen, 1981: 52). Sen's figures are contested. Goswami (1990) estimated these losses 32 per cent relative to the foregoing year and 19 per cent relative to the average of the last five years. In Box 6.6 I used Uppal's (1884) data.

Box 6.6. Some quantitative considerations on the Bengal rice harvests in 1942.
Bengal statistics are confusing, to say the least[474]. Here I use data provided by Uppal (1984).

Bengal rice area in the years 1938-40 was 25 million acres or ~10.3 million hectares, yielding ~0.38 tonnes per acre or 0.964 tonnes/ha (*ibid.*: 73). Projecting these figures on 1942 we find a total rice (paddy) yield of ~10 million tonnes, disregarding any mishaps. This figure approximates that quoted for total rice production over 1942, 10.8 million tonnes (*ibid.*: 73). Presumably, the cyclone and disease effects were incorporated in this figure. Bengal needed about 10 million tonnes per year (*ibid.*: 58). The total amount of rice available in 1942 was ~10.1 million tonnes (*ibid.*: 73), ranking 1942 among the three top years (all with >10 million tonnes) out of the last fifteen years. The population of Bengal in 1941 amounted to 60.3 million (*ibid.*: 54). Using this figure for 1942 the result is ~166 kg rice (paddy) per person per year, including infants, children, adults, and elderly but excluding the military and the hundreds of thousands of refugees from Burma. The 166 kg per person per year amount to 454 gram paddy per person per day and 0.7 x 454 = 319 gram of milled rice per person per day, or ~1100 Calories per person per day. Considering the sizeable imports of wheat, the 1100 Calories seem neither too good nor too bad.

The area damaged by the cyclone was estimated at 3200 square miles (*ibid.*: 67), equal to ~830.000 ha. Assuming this area to incorporate open waters, roads, built areas and non-rice crops, I guess that some 400,000 ha of paddy fields were lost, or 4 per cent of the total Bengal rice area. One source (*ibid.*: 67) stated the rice diseases caused more damage than the cyclone, say between 4 and 8 per cent loss of paddy for all of Bengal.

This relatively modest loss estimate is very different from the percentages given by Padmanabhan (1973), who probably selected the two worst affected trial fields among the many variety trials scattered over Bengal with the intent to show the extremes of loss due to brown spot. It is plausible that there were indeed many variety trials, as also suggested by Tauger (2003), and that the data from the not-mentioned trials did not serve Padmanabhan's purpose.

With the mental reservation that the Bengal statistics I could examine so far are rather confusing, I conclude that both cyclone and disease must have been disastrous in certain areas, but that neither cyclone nor disease can be held responsible for the wide-spread famine of 1943 following the unsatisfactory but not too bad harvests of 1942.

Unfortunately, I haven't found an independent testimony. Brennan (1988) mentions crop failure, in part due to July rains that damaged the *aus* crop, a minor crop preceeding the *aman* crop and harvested in September-October. The *aus* crop may have provided massive brown spot inoculum for the *aman* crop. Other mishaps in 1942, exceptionally high floods and severe cyclone damage in parts of Bengal, contributed to the crop failure[475]. Tauger (2003) brushed up the importance of the disease, using Padmanabhan's data, in an unwarranted extrapolation, without offering new information.

6.5.3. Food availability and purchasing power

At first sight the present case is a straightforward example of a famine due to a plant pathogen. However, there is more to it. In the exchange of telegrams and letters between the British and the British Indian Governments cyclone and flood damage had been mentioned (Mansergh, 1973: 437) but not disease damage. Amarthya Sen (1981), Nobel Price Laureate in economics, being in Calcutta as a 6-year old boy was another witness of the misery. He saw people dying in front of well-stocked shops. Sen made a cool socio-economic *ex post facto* analysis. The total amount of food in Bengal, 1942, had never been so large, he wrote, and 'on average' it would have been sufficient to feed the population, even though imports from Burma and from other Indian states had been cut off due to war conditions.

World War II raged from 1939 to1945. The Japanese troops, having conquered Burma, were threatening India and the British Government of India had to think of war, displacement of troops, and supplying the war machine. Thanks to all these unusual activities a war economy was booming. A rampant inflation was fed by printing fresh bank notes. Rice prices sky-rocketed. The price index in August 1943 was 536 with December 1941 as 100 (Sen, 1981: 66). At the same time, disasters of different nature occurred, cyclone damage with high floods, local drought affecting the jute production, and falling jute prices. In addition, local society was disrupted by political strife, the 'Quit India Movement' in late 1942 (Brennan, 1998: 548).

Sen (1981: 52) and Greenough (1982) mentioned 'a plant disease' only in passing, rather for completeness sake than as an explanatory variable. Both stated that the famine was 'man-made'. By error of judgement and serious mismanagement of the Bengal government, not to mention corruption, the Bengal authorities could not prevent the famine. Measures taken by the Administration to mitigate the famine had contrary effects, storage of rice and hoarding, and price boosts. In fact, enough rice was available but the working class and the small farmers could no longer pay the price. Calculations (Box 6.6) suggest that neither cyclone nor brown spot disease could have caused the Bengal-wide famine of 1943.

The Bengal social system with gives and takes for landowners and their dependents, farmers and otherwise, broke down completely because stocks and funds of the landowners dried up. Landlords, no longer able or willing to feed their tenants, had to dismiss dependents and these added to the hungry mass of landless labourers who, without a source of income, migrated in search of food. Social disruption helps to explain the multitude of the landless, roaming about, begging for food, starving and dying.

Indeed, the Bengal Famine was 'man-made'[476]. Besides the incompetence of the Bengali Government[477] and the corruption within the administration, besides the unrest caused by the war and by the national politics with forward-casting shadows of Partition (1947) and Independence (1950), Greenough (1982) described another and underexposed man-made effect. The time-honoured Bengali social system had cracked under pressure of imported capitalistic attitudes. Now it broke down. Under unbearable stress even individual families were broken up. So the 'man-made' effect occurred at two levels, the provincial government at the top and the landowners and tenant families at the base. Starvation reached its peak in November, 1943, at about the time the bumper crop of 1943 was being harvested.

Landless labourers and small peasants, poor anyhow, who had lost their crop were left without purchasing power. They could not buy whatever was available. India had a long tradition in dealing with famines, a governmental Famine Code stemming from the 19th century. Famines in one or another state belonged to the facts of Indian life. India was so large that usually food could be brought in from elsewhere by truck and train, but it had to be paid. 'Work for food' programs were not unusual.

The British Government, struggling for the survival of Great Britain, could not spend much attention to India. The Viceroy of India, Lord Wavell wrote in a letter to Prime-Minister Churchill dated 24 October 1944 'the vital problems of India are being treated by His Majesty's Government with neglect, even sometimes with hostility and contempt' (Sen, 1981: 97). The situation was complicated because of the world-wide war and because in India political events threw their shadows ahead, strife between Muslim and Hindu factions, struggle for independence. The Partition between largely Hindu India and primarily Muslim Pakistan[478] dates from 1947. The famine itself became part of the ammunition in the psychological warfare between the nationalists and the British, which led to the independence of India in 1950.

6.5.4. Conclusions

1. The dark and wet fall in Bengal, 1942, may have reduced the rice yields and may have sensitised the rice crop so that an epidemic of brown spot on rice could develop.
2. Indeed there was a serious outbreak of brown spot in the *aman* crop of 1942, but its severity, extent, and impact on yield are not known. Dull weather, floods and cyclone damage may have been more important than plant disease.
3. The brown spot epidemic of late 1942, preceding and associated with the Great Bengal Famine of 1943, was not its immediate cause.
4. The real cause was poor governance. The famine was typically man-made.

6.6. An afterthought

The epidemiologist has the task to solve the riddles of an epidemic, caused by one pathogen with one vector on one crop. Though fascinatingly complex in detail the issue itself is straightforward, a one-to-one-to-one problem. The historian will try to describe and explain the 'human condition', or some aspect thereof, finding a grand picture by piecing together innumerable bits of evidence, and putting aside a hundred times more.

Whereas the epidemiologist hardly touches the human condition the historian has, often, little eye for agricultural detail, let alone for phytopathological finesse. The historical epidemiologist may try to link biological bits and pieces, that shape an epidemic, to the human condition that manifests itself by phenomena such as famine, revolution, or war. The historical epidemiologist should find his way through two scientific areas to collect his own scraps of evidence, having put on his own blinkers to avoid distraction by interesting side-issues. The historian can reduce a plant disease epidemic to a footnote under a dramatic episode of human history, and he has obvious difficulties to handle something that did not happen, a non-event.

Remarkably, a historian – Matossian – unearthed the role of a plant disease in a drama of decisive historical importance, the Great French Revolution of 1789, a supportive role rather than the leading part (§6.2). Historians have little to tell about the military campaign of 1722 that may have failed because of a plant disease (§6.3). Once I believed that a plant disease was a major player in the Russian Famine of 1932-33 (§6.4), but now I think it was an unfortunate coincidence of an epidemic with the industrialisation of Russian agriculture. The general neglect delayed the ripening of the wheat and enhanced the epidemic that might have added to the food shortage. Similarly, a plant pathologist over-emphasised the contribution of a plant disease to the Bengal Famine of 1943 (§6.5), whereas an economist reduced its importance to a few side remarks only.

In all four cases the agricultural data are too poor to pass a final judgement. In all four cases poor and sometimes even criminal governance enhanced the impact of plant disease. Obviously, the rulers addressed in the introductory quotations failed miserably.

7. Postscriptum

The preceding chapters dealt with plant disease epidemics and tried to connect agriculture and its frequent failures to human affairs at large. Poor governance often aggravated the nefarious effects of plant disease epidemics and good governance seldom mitigated them. From utter helplessness, as in Chapter 1, to the refined but still not perfect crop protection of today mankind covered much mileage along the road from poor to good governance, promoting agricultural science and harnessing social science for the fight against hunger. The end of the road is, however, not yet in sight.

'... the grand essential to the exercise of royal government, a heart on the part of the Sovereign impatient of the sufferings of the people, and eager to protect them and make them happy ...' (conversation with King Hsüan).

Mencius (Mäng-Tsze) ~372-289 BC (Legge, 1970)

7.1. Food and politics

In the Western World of today, around the year 2000, food and agriculture have become non-issues. Food is always available at the supermarket. The consumer's main concern is the length of the queue at the checkout.

Some 160 years ago, say 1850, food was a capital issue. The well-being and even the size of a population depended on its food production, in a Malthusian way. Crops were threatened by all kinds of natural phenomena among which bad weather – either too hot or to cold, too dry or too wet – , volcanic eruptions, infestation by weeds, and outbreaks of pests and diseases.

To these natural phenomena should be added the man-made phenomena, the burden of taxation, extortion, hunting rights, war, and poor governance. Food supply was insecure. Grain shipping overseas – in sailing vessels – was relatively well-organised but grain transportation overland – in bullock-drawn carts over muddy roads – was rather un-organised, awkward and risky. Grain storage was far from perfect. Hoarding and speculation were common in times of want, notwithstanding frequent prohibitions.

When scarcity and dearth reigned, the rank and file became rebellious. They reproached the authorities of town and state to take inadequate care of the commoner and the poor. Local risings were frequent, often with women in the forefront. Risings could lead to revolts, and revolts to revolutions. In the past there were obvious links between adverse natural conditions, economics, and politics.

7.2. Six capita

Towards the end of a career devoted to the study of plant disease epidemics I became intrigued by these links, choosing epidemics as one species of natural phenomena causing disaster. The *capita selecta* of this book are tied together by a curiosity-driven, personal choice. Plant disease epidemics come and go, as individuals *sui generis* (Gäumann, 1946). What is their place in the course of human events? Imperceptibly a structure imposed itself.

Caput 1 reaches back to my high-school years and my fascination with the classics, Biblical, Greek, and Latin. In the pre-scientific era, when calamities could not be understood in the way we pretend to do now, the super-natural was nearby. No clear borderline between the natural and the super-natural was seen. Fear and despair, hope and ambition were made manageable by religion. Preachers interpreted calamities, including those due to rust and blight, as the wrath of the gods shattered over mankind in punishment of disobedience. Where no remedy was available people wanted to do at least something and turned to magic.

Caput 2 is a late echo of my most creative years, when imaginative principals stimulated me to prepare a doctoral thesis (Zadoks, 1961) on yellow rust of wheat, covering half of Europe in space and half a century in time. The calamitous year 1846 was a rust year. Wheat suffered from an un-identified rust, possibly yellow rust, but the real problem was yellow rust on rye. The epidemic was reconstructed, using many tiny bits of information, but the eastern wing of the epidemic remained a bit nebulous. The effects of drought and rust, an unusual combination, could not be sorted out satisfactorily. Their economic and social effects were overshadowed by those of potato late blight.

As a career epidemiologist I could not resist the temptation to study potato late blight though little came out of it. Caput 3 reflects my interest in the 'potato murrain' during the tumultuous years 1845-1848.

Caput 3. Mencius, the Chinese sage quoted above, saw the Good Prince as key to human welfare, based on a stable agriculture. Under a wise ruler, he believed, no harm will touch the crops. This is, unfortunately, too good to be true. The 'potato murrain' appeared out of the blue in 1844, hitting hard in 1845 and 1846. No Good Prince could not have avoided the blow but, in the modern disguise of Good Governance, he could have done more to mitigate its nefarious effects on human affairs. Poor Governance ruined France in 1848. The Prince, King Louis Philippe d'Orléans, lost his throne. The governments of Belgium and the Netherlands responded immediately to the emergency of 1845 and, according to the standards of the time, adequately. Belgium, then a new and energetic nation, went one step further in meeting the needs of its people than the Netherlands, but neither country could prevent hunger and distress. We cannot accuse these countries of poor governance but neither can we praise them for excellence in the sense of Mencius.

The Capita 2 and 3 form a pair, as do the Capita 4 and 5.

Caput 4. Growing into adolescence during World War II, half-aware of scarcity and famine in parts of the Netherlands, I became dedicated to agriculture. From Word War II my thoughts went back to World War I. By chance I found out about my grandfather, then an Alderman of Amsterdam, who had to face angry women demanding potatoes, not just food but potatoes. What happened in Holland during that gruesome period, and what in neighbouring Germany?

Caput 5 sketches the complexities of feeding Germany, a country at war, successful in military campaigns but isolated from the world. Though potato late blight was important in 1916, the blight was but one nasty event among several other agricultural mishaps. Mismanagement of the food chain contributed to the distress of the people at home and in the end also of the soldiers at the front. What was worse, the poor governance by Germany's rulers or the arrogant ignorance of the military in matters of food?

Caput 6. A plant disease epidemiologist, focused too much on 'his' epidemic, runs the risk of becoming myopic. Linking the 1932-33 Russian Famine to the black stem rust epidemic of 1932 I fell into that trap. A prominent Indian scientist was similarly entrapped when connecting the 1942 brown spot epidemic on rice to the Bengal Famine of 1943. Historians and economists reduced the Russian rust epidemic to a footnote and the Bengal brown spot epidemic to a few casual remarks. Poor Governance in the latter case and Criminal Governance in the former caused human tragedies with millions of deaths. Poor Governance prepared the ground for the French Revolution of 1789. The exasperation of the rural poor, possibly enhanced by ergotism, culminated in *la Grande Peur*, the panic of 1789. The Prince, King Louis XVI, tried to be good but his efforts were too late and too little. He lost his throne and his head. The Great French Revolution changed the Western World profoundly.

7.3. Politics and food

Of old, farmers were practically helpless against the calamities that visited them time and again, be it drought or flood, volcanic activity, war, or excessive exploitation. The 'Good Prince' supported his farmers, was trusted by his farmers, maintained law and order, organised public works such as roads, irrigation and drainage systems, and – in exchange – took his share of the produce with moderation. Good Princes were rare, but several are on record that opened their stores to people in need. Joseph, Viceroy of Egypt, is the shining example of a Prince with foresight (Genesis 41 and 47).

The science of crop protection was of little avail during the emergencies discussed in this volume but farmers were not totally helpless. The lore and science of crop husbandry was developed gradually. In the 18th century England took the lead. Early in the 19th century agricultural science was remarkably well established in France. Later in the 19th century Germany took over. New varieties of crops were selected, consciously or unconsciously. Chemical control came into being in the course of the 19th century. Academics, government-paid, took the lead in developing agricultural science. Toward the end of the 19th century governments became interested and initiated agricultural research stations, a great step on the road toward Good Governance (Koeman & Zadoks, 1999).

In the 20th century governments gradually came to recognise the power of agricultural science, including crop protection science. Mankind learned – slowly – to control plant disease epidemics by good husbandry, preventing them by genetics, and fighting them by chemistry. The ill effects of such epidemics on humans can be attenuated by various social and economic measures. Strategic reserves of grain were set up that contribute to the resilience of societies. In the case of an emergency, international food-streams can be re-directed. Good Governance has made progress indeed. Around the year 2000, nonetheless, many are hungry or under-nourished. More work needs to be done. This task is, however, largely beyond the reach of epidemiology.

7.4. Final comments

Being a natural scientist I am not familiar with the rigorous methods of the historian who, supposedly, digs deep into manuscripts, minutes of meetings, hand-written letters, throw-away notes on scraps of paper, and contemporaneous newspapers. Formal publications were the basic material of the studies presented here. Nonetheless, these provide new information, unexpected combinations, or fresh criticism. A consistent overall view of the historical impact of plant disease epidemics was beyond reach but a start has been made.

The results I offer to my family, friends and relations at the occasion of my eightieth birthday with a feeling of gratitude that I have been allowed to work in and for agriculture, digging deep into the challenging field of plant disease epidemiology, and sprawling out over the areas of research, education, extension, administration and consultancy, within the narrow confines of the Netherlands but also far beyond, in Europe, and in the world at large.

Acknowledgements

The kind invitation by the Swedish Agricultural University at Uppsala, 2005, to accept an Honorary Doctorate from the Faculty of Agricultural Sciences, was at the origin of this book. I was requested to present my reflections on plant disease epidemiology to an educated public unfamiliar with the details of my trade. The resulting two lectures now form the core of the Chapters 3 and 6. Fascinated by the unexpected perspectives I continued my investigations.

My wife, Mrs J.C. van Heuven created a favourable environment to ruminate the subject matter and write. She was an invaluable source of criticism and stimulation.

Professor P.C. Struik (Wageningen University) carefully read the manuscripts of several chapters and provided valuable suggestions. Professor F.P.M. Govers (Wageningen University) critically contributed to the Chapters 3, 4, and 5 on *Phytophthora infestans*.

Chapter 1. Several spokesmen participated in their youth in the field processions of the Rogation Days, around World War II (1939-1945). Mr. B. Kruseman (Zeiss Planetarium, The Hague) advised me in 1967 on the astronomical implications of celestial dates. Dr. L.B. van der Meer (Leiden State University, the Netherlands) contributed by constructive criticism and valuable suggestions. Professor Amos Dinoor, Hebrew University (Rehovot), gave useful information. Mr. A.P. Zadoks critically read earlier versions of the manuscript.

Chapter 2. Professor Jonathan Yuen (Swedish University of Agricultural Sciences, Uppsala) kindly provided information on and translation of Swedish texts.

Chapter 3. Thanks are due to Dr. S. Savary (France) for many useful suggestions. Professor Jonathan Yuen (Swedish University of Agricultural Sciences, Uppsala) contributed valuable information. Dr Pamela Anderson (CIP, Peru) showed a most stimulating interest in the subject matter of this chapter.

This book could not have been written without the assistance of many libraries. The Dutch libraries contain a wealth of old journals and books that can be found by means of the computerised PICARTA system. The organisation of these libraries with their many sub-libraries, spread out over numerous addresses, is somewhat confusing, and so that I mention only some major libraries:
- The Library of Wageningen University and Research Center (WUR), its Special Collections, and its former Crop Protection Centre Library in particular.
- The University Libraries of Amsterdam, Leiden, and Groningen.
- The Royal Library at The Hague.
- The libraries of the Netherlands Entomological Society at Amsterdam, the Amsterdam Zoo 'Artis Natura Magister', the National Herbarium at Leiden, the Municipal Archive of Amsterdam, the Peace Palace at the Hague, and Teylers Museum at Haarlem.

Publications by J.C. Zadoks relating to the history of plant disease

1961. Yellow rust on wheat, studies in epidemiology and physiologic specialization. Tijdschrift over Plantenziekten (Netherlands Journal of Plant Pathology) 67: 69-256.

1965. Epidemiology of wheat rusts in Europe. FAO Plant Protection Bull. 13: 97-108.

1969. Hogg, W.H., Hounam, C.E., Mallik, A.K. & Z. – Meteorological factors affecting the epidemiology of wheat rusts. World Meteorological Organization, Techn. Note No. 99, Geneva, Switzerland: 143 pp.

1976. Z & Koster, L.M. – A historical survey of botanical epidemiology. A sketch of the development of ideas in ecological phytopathology. Meded. Landbouwhogeschool, Wageningen 76-12: 56 pp.

1981. Mr. DUHAMEL's 1728 treatise on the violet root rot of saffron crocus: 'Physical explanation of a disease that perishes several plants in the Gastinois, and Saffron in particular'. Meded. Landbouwhogeschool, Wageningen 81-7: 31 pp.

1984. Cereal rusts, dogs, and stars in Antiquity. Garcia de Orta, Ser. Est. Agron. Lisboa 9 (1982): 13-29.

1984. A quarter century of disease warning, 1958-1983. Plant Disease 68: 352-355.

1985. Z & Bouwman, J.J. – Epidemiology in Europe, pp. 329-369. In: Roelfs, A.P., Bushnell, W.R. (Eds.) The cereal rusts, Vol. II. Orlando, Academic Press. ca 600 pp.

1988. Twenty five years of botanical epidemiology. Philosophical Transactions Royal Society London B 321: 377-387.

1991. A hundred and more years of plant protection in the Netherlands. Netherlands Journal of Plant Pathology 97: 3-24.

1992. The partial past. Comments on the history of thinking about resistance of plants against insects, nematodes, fungi, and other harmful agents, pp. 11-22. In: Th. Jacobs and J.E. Parlevliet (Eds.), Durability of disease resistance. Kluwer Academic Publishers, Dordrecht NL. 375 pp.

1994. Polyetic epidemics by plan or contingency, pp. 139-145. In: R.A. Leigh & A.E. Johnston: Long-term experiments in agricultural and ecological sciences. Wallingford, CAB International. 428 pp.

1995. A short history of IPPC. Abstracts XIII International Plant Protection Congress, The Hague, The Netherlands, 2-7 July 1995.

1999. Koeman, J.H. & Z. – History and future of plant protection policy, from ancient times to WTO-SPS, pp 21-48 and 225-231. In: G. Meester, R.D. Woittiez & A. de Zeeuw (Eds.) Plants and politics. On the occasion of the centenary of the Netherlands' Plant Protection Service. Wageningen, Wageningen Pers.

1999. Reflections on space, time and diversity. Annual Review of Phytopathology 37: 1-17.

2000. Z & Waibel. H. – From pesticides to genetically modified plants: history, economics and politics. Netherlands Journal of Agricultural Science 48: 125-149.

2001. Plant disease epidemiology in the twentieth century: a picture by means of selected controversies. Plant Disease 85: 808-816.

2002. Zwankhuizen, M.J. & Z. *Phytophthora infestans*'s 10-year truce with Holland: a long-term analysis of potato late-blight epidemics in the Netherlands. Plant Pathology 51: 413-423.

2003. Fifty years of crop protection, 1950-2000. Netherlands Journal of Agricultural Science 50: 181-193.

2003. Two wheat septorias; two emerging diseases from the past, pp. 1-12. In: G.H.J. Kema, M. Van Ginkel & M. Harrabi (Eds.): Global insights into the *Septoria* and *Stagonospora* diseases of cereals. Proceedings of the 6th International Symposium on *Septoria* and *Stagonospora* Diseases of Cereals, December 8-12, 2003, Tunis, Tunesia.

2007. Fox and fire – a rusty riddle. Latomus (Brussels) 66: 3-9.

2007. Z & Ariena H.C. Van Bruggen – Johanna Westerdijk (1883-1961). The grand lady of Dutch plant pathology, pp 155-167. In: J.B. Ristaino (Ed.) Pioneering women in plant pathology. Minnesota, APS Press.

2008. The potato murrain on the European Continent and the revolutions of 1848. Potato Research 51: 5-45.

References

Abel, W. – 1974. Massenarmut und Hungerkrisen im voridustriellen Europa: Versuch einer Synopsis. Hamburg, Parey. (*Mass poverty and hunger crises in pre-industrial Europe: Attempt at a synopsis*).

Achilli, A., Olivieri, A., Pala, M., Metspalu, E., Fornarino, S., Battaglia, V., Accetturo, M., Kutuev, I., Khusnutdinova, E., Pennarun, E., Cerutti, N., Di Gaetano, C., Crobu, F., Palli, D., Matullo, G., Santachiara-Benerecetti, A.S., Cavalli-Sforza, L.L., Semino, O., Villems, R., Bandelt, H.J., Piazza, A., Torroni, A. – 2007. Mitochondrial DNA variation of modern Tuscans supports the Near East origin of Etruscans. American Journal Human Genetics 80: 759-768.

Ackersdijck, J., Alstorphius Grevelink, P.W., de Bosch Kemper, J., den Tex, C.A., de Meester, G.A., Elink Sterk jr., A., Mees, W.C., Portielje, D.A., Vissering, S., *et al.* – 1850. Staatkundig en staathuishoudkundig jaarboekje voor 1850. 2^e jaargang. Amsterdam, Müller.

Ackersdijck, J., *et al.* – 1851. Staatkundig en staathuishoudkundig jaarboekje voor 1851. 3^e jaargang. Amsterdam, Müller.

Ackersdijck, J. *et.al.* – 1852. Staatkundig en staathuishoudkundig jaarboekje voor 1852. 4^e jaargang. Amsterdam, Müller.

Adanson, M. – 1763. Familles des plantes. I. Partie. Paris, Vincent. (*Families of plants*).

Aereboe, F. – 1927. Der Einfluss des Krieges af die landwirtschaftliche Produktion in Deutschland. Stuttgart, Deutsche Verlags-Anstalt. (*The influence of the war on the agricultural production in Germany*).

Agulhon, M., Désert, G., Specklin, R. – 1976. Histoire de la France rurale de 1789 à 1914. Paris, Seuil. (*History of rural France from 1789 to 1914*).

Albert, W. – 1845. Authentische Thatsachen über die Verjüngung der Kartoffeln aus Samenkörnern, mit Berücksichtigung der unter denselben jetzt herrschenden Krankheiten. Magdeburg, Bänsch. (*Authentic facts about the rejuvenation of the potatoes from seed grains, with consideration of the diseases now reigning among them*). From Oekonomische Neuigkeiten 1847, p. 496.

Algemeen Verslag wegens den Staat van den Landbouw in het Koninkrijk der Nederlanden – 1845, 1846, 1847. Haarlem, Loosjes. (*General Report on the state of agriculture in the Kingdom of the Netherlands*).

Allgemeine Zeitung, Augsburg edition. Cotta, Stuttgart. (*General Newspaper*).

André, R., Pereira-Roque, J. – 1974. La démographie de la Belgique au XIXe siècle. Brussels, Éditions de l'Université de Bruxelles. (*The demography of Belgium in the 19th century*)

Anikster, Y., Wahl, I. – 1979. Coevolution of the rust fungi on Gramineae and Liliaceae and their hosts. Annual Review of Phytopathology 17: 367-403.

Annual Reports – 1903/1930. Verslagen en Mededeelingen van de Directie van den Landbouw. The Hague, van Langenhuysen. (*Reports and communications of the Directorate of Agriculture*).

Anonymous – 1789. Nieuwe Nederlandsche Jaarboeken, of vervolg der merkwaardigste geschiedenissen, die voorgevallen zijn in de Vereenigde Provinciën, de Generaliteitslanden, en de Volksplantingen van den Staat. 24ste deel. Leiden, Van der Eyk & Vijgh. (*New Dutch Yearbooks ... or continuation of the most curious histories ..., Vol. 24*).

Anonymous – 1846a. Ernteberichte vom Jahre 1845. Oekonomische Neuigkeiten und Verhandlungen 71: 193-195, 204-206. (*Harvest notes from the year 1845*).

Anonymous – 1846b. Bederf in de Rogge. Mededeelingen en handelingen van de Geldersche Maatschappij van Landbouw 1: 45-59. (*Decay in the rye*).

Anonymous – 1846c. Bederf in de Aren der Rogge. Mededeelingen en handelingen van de Geldersche Maatschappij van Landbouw 1: 97-101. (*Decay in the ears of the rye*).

Anonymous – 1846d. De Aardappelziekte. Mededeelingen en handelingen van de Geldersche Maatschappij van Landbouw 1: 209-241. (*The potato disease*).

References

Anonymous -1846/7. Tijdschrift ter Bevordering van de Nijverheid 15 (1846) 330-376 and 16 (1847) 323-345. (*No title*).

Anonymous – 1847a. Übersicht der Erndte-Erträge in der Preuszischen Monarchie im Jahre 1846. Annalen der Landwirthschaft in den Königlich Preussischen Staaten, Jahrgang 5 Band 10: 400-401. (*Overview of harvest yields in the Prussian Monarchy in the year 1846*).

Anonymous – 1847b. Drei Berichte des Directoriums des landwirthschaftlichen Haupt-Vereins zu Münster, die Rostkrankheit des Roggens betreffend. Annalen der Landwirthschaft in den Königlich Preussischen Staaten, Jahrgang X Band Y: 179-181. (*Three notes from the Directorate of the Agricultural Main Society at Münster, on the rust disease of rye*).

Anonymous – 1847c. No title. Landes Oeconomie-Collegium, January 4th. Annalen der Landwirthschaft 5 (Vol. 10): 389-401.

Anonymous – 1847d. Über die Kartoffelkrankheit in Esthland im Jahre 1846. Oekonomische Neuigkeiten und Verhandlungen 72: 695. (*On the potato blight in Estonia in the year 1846*).

Anonymous – 1848a. Die Hungerpest in Oberschlesien: Beleuchtung oberschlesischer und preussischer Zustände. Mannheim, Heinrich Hoff. (*The hunger pest in Upper Silesia: elucidation of Upper Silesian and Prussian conditions*).

Anonymous – 1848b. Die Erndte-Erträge in der Preuszischen Monarchie im Jahre 1847 betreffend. Annalen der Landwirthschaft 9 (Vol. 11): 370-385. (*The harvest yields in the Prussian Monarchy in the year 1847*).

Anonymous – 1849. Staatkundig en staathuishoudkundig jaarboekje voor 1848. Amsterdam, Müller. (*Economic and political yearbook for 1848*).

Anonymous – 1850a. Statistiek van den handel en scheepvaart 3: 34vv. (*Statistics of trade and shipping*).

Anonymous – 1850b. Statistique de la Belgique. Agriculture. Recensement général (15 Octobre 1846). I-V. Bruxelles. (*Statistics of Belgium. Agriculture. General census*).

Anonymous – 1910. Beschrijving van de aardappelcultuur in Nederland. Verslagen en Mededeelingen van de Directie van den Landbouw No 3: I-XXX, 1-183. (*Description of potato cultivation in the Netherlands*).

Anonymous – 1914. Mitteilungen der Wiener Pflanzenschutzstation. Abstract in Zeitschrift für Pflanzenkrankeiten: 463. (*Communications of the Viennese station of plant protection*).

Anonymous – 1916. No title. Mitteilungen aus den Kaiserlichen Anstalt für Land- und Forstwirtschaft 10: 59-61.

Anonymous – 1917. Verslag 1917. Gemeenteblad van Amsterdam Vol. I-2, p. 859-868; Vol. II-2 p. 1265-1479. (*Amsterdam City Council report on 1917*).

Anonymous – 1919. No title. Mitteilungen aus den Kaiserlichen Anstalt für Land- und Forstwirtschaft 17: 17, 47-49.

Anonymous – 1996. Écritures en Méditerranée, 2nd Ed. Tunis, Alif. (*Scripts in the Mediterranean area*).

Appel, O. – 1906a. Über die Blattrollkrankheit der Kartoffel. Mitteilungen aus der Kaiserlichen Biologischen Anstalt für Land- und Forstwirtschaft 2: 10-11. (*On the leaf roll disease of potatoes*).

Appel, O. – 1906b. Neuere Untersuchungen über Kartoffel- und Tomaten-erkrankungen. Jahresbericht der Vereinigung der Vertreter der angewandten Botanik (2er Jahrgang 1904/5) 3: 122-136. (*Newer studies on potato and tomato diseases*).

Appel, O. – 1917. Die Rhizoctonia Krankheit der Kartoffel. Deutsche Landwirtschaftliche Presse p. 499. (*The Rhizoctonia disease of the potato*).

Appel, O. – 1928. Handbuch der Pflanzenkrankheiten 2er Band. Die pflanzliche Parasiten, 1er Teil. 5e Aufl. Berlin, Parey. (*Handbook of plant diseases, the vegetable parasites*)

Appel, O. – 1931. Getreidekrankheiten. Berlin, Parey (pocket atlas). (*Diseases of cereals*)

On the political economy of plant disease epidemics

Armengaud, A. – 1961. Les populations de l'Est-Aquitain au début de l'époque contemporaine. Recherches sur une région moins développée (vers 1845 – vers 1871). Paris, Mouton. (*The populations of East Aquitania at the beginning of the contemporaneous era. Studies on a less developed region (~1845 - ~1871)*).

Armengaud, A. – 1971. La population Française au XIXe siècle. Paris, Presses Universtaires de France. (*The French population in the 19th century*).

Audoin-Rouzeau, S. – 2003. Kinder und Jugenliche, p. 135-141. In: Hirschfeld, G., Krumeich, G., Renz, I. (Eds.), Enzyklopädie Erster Weltkrieg. Paderborn, Schöningh. (*Children and younsters*).

Azzi, G. – 1930. Le climat du blé dans le monde. Rome, Institut International d'Agriculture. (*The climate of wheat in the world*).

Baars, C. – 1973. De geschiedenis van de landbouw in de Beijerlanden. Wageningen, Pudoc. (*The history of agriculture in the Beijerlanden, Netherlands*).

Badinter, E. – 1978. Les 'Remontrances' de Malesherbes 1771-1775. Paris, Union Générale d'Éditions. (*The remonstrances of Malesherbes 1771-1775*).

Bailey, S.D. – 1945. Post mortem on the Bengal famine. Far Eastern Survey 14: 373-374.

Bain, C., Bernard, J.-L., Fougeroux, A. – 1995. Protection des cultures et travail des hommes. Paris, Le Carrousel. (*Crop protection and human effort*).

Bakels, H. – 1946. Bijbelsch woordenboek, 3rd Ed. Amsterdam, Wereldbibliotheek. (*Biblical dictionary*).

Banks, J. – 1805. On the blight of corn. Annex to Curtis (1805).

Barger, G. – 1931. Ergot and ergotism, A monograph based on the Dohme Lectures delivered in Johns Hopkins University, Baltimore. London, Gurney & Jackson.

Bergman, M. – 1967. The potato blight in the Netherlands and its social consequences (1845-1847). International Review of Social History 12: 390-431.

Bergsma, C.A. – 1845. De aardappel-epidemie in Nederland in den jare 1845. Utrecht, Van Terveen. (*The potato-epidemic in the Netherlands in the year 1845*).

Berkeley, M.J. – 1846. Observations, botanical and physiological, on the potato murrain. Journal of the Horticultural Society of London. 1: 9-34. Reprinted in Phytopathological Classics 8 (1948) 1-108.

Berkeley, M.J. – 1854/7. Vegetable Pathology. Gardeners' Chronicle and Agricultural Gazette, 1854-1857. Selections as reprinted in Phytopathological Classics 8 (1948). Quotations give the original year of publication, Berkeley's paragraph number (§), and the page number as in Phytopathological Classics.

Berkeley, M.J. – 1860. Outlines of Britsh fungology. London, Lovell Reeve.

Berthold, R. – 1974. Zur Entwicklung der Deutschen Agrarproduktion und der Ernährungswirtschaft zwischen 1907 und 1925. Jahrbuch für Wirtschaftsgeschichte 1 (4): 83-111. (*On the development of the German agricultural production and the food business between 1907 and 1925*).

Besteman, R. – 2007. De aardappelziekte: Hide and seek in de polder. Gewasbescherming 38: 207-210. (*The potato late blight: hide and seek in the polder*).

Bickel, W. – 1947. Bevölkerungsgeschichte und Bevölkerungspolitik der Schweiz seit dem Ausgang des Mittelalters. Zürich, Guteberg. (*Population history and population policy of Switzerland since the end of the middle Ages*).

Bieleman, J. – 1987. Boeren op het Drentse zand 1600-1910. Een nieuwe visie op de 'oude landbouw'. Wageningen, A.A.G. Bijdragen 29. (*Farming on the sands of Drente 1600-1910. A fresh view on the 'old agriculture'*).

Bieleman, J. – 1992. Geschiedenis van de landbouw in Nederland 1500-1950. Meppel, Boom. (*History of agriculture in the Netherlands 1500-1950*).

Bieleman, J., Van Otterloo, A.H. – 2000. Techniek in Nederland in de twintigste eeuw. III. Landbouw en Voeding. Zutphen, Walburg Pers. (*Technology in the Netherlands in the 20th century. III. Agriculture and nutrition*).

Billiard, R. – 1928. L'Agriculture dans l'antiquité d'après les Géorgiques de Virgile, Paris, Boccard. (*Agriculture in antiquity after Virgil's Georgica*).

Bjerkander, C. – 1794. Anmärkningar, vid hvilken tid trän och rötter fingo mogen frukt och frön innevarande år. Vet. Akad. Nya Handlingar, Stockholm. (*Observations on the time when the trees and roots get ripe fruits and seeds in the current year*).

Blaive, F. – 1995. Le rituel romain des Robigalia et le sacrifice du chien dans le monde indo-européen. Latomus (Brussels) 54: 279-289. (*The Roman ritual of the Robigalia and the dog sacrifice in the indo-european world*).

Blanchard, XX. & Perret, XX. – 1917. No title. CR Académie d'Agriculture de France 3: 894-895.

Blok, D.P. – 1977. Algemene Geschiedenis der Nederlanden. 12. Nieuwste tijd: Nederland en België: 1840-1914. Haarlem, Fibula. (*General history of the Netherlands*).

Blum, J. – 1948. Noble landowners and agriculture in Austria, 1815-1848. A study in the origins of the peasant emancipation of 1848. Baltimore, Johns Hopkins Press.

Blume, XX. – 1845. No title. Algemeen Handelsblad of 15 August 1845, N°4287: 4.

Bolognese-Leuchtenmüller, B. – 1978. Wirtschafts- und Sozialstatistik Österreich-Ungarns. I. Bevölkerungsentwicklung und Berufsstruktur, Gesundheits- und Fürsorgewesen in Österreich 1750-1918. Vienna, Verlag für Geschichte und Politik. (*Economic and social statistics of Austria-Hungary ...*).

Bömer, F. – 1956. Die römischen Ernteopfer und die Fuchse im Philisterland. Wiener Studien: Zeitschrift für classische Philologie 69: 372-384. (*The Roman harvest sacrifices and the foxes in the land of the Philistines*).

Bömer, F. – 1958. P. Ovidius Naso. Die Fasten. Vol. II. Heidelberg, Winter. (*The Fasti*).

Bonjean, J. – 1846. Monographie de la pomme de terre. Envisagée dans ses rapports agricoles, scientifiques et industriels et comprenant l'histoire générale de la maladie des pommes de terre en 1845. Paris, Baillière. (*Monography of the potato, considered in its agronomic, scientific and industrial relations and including the general history of the potato disease in 1845*).

Bouchard, G. – 1972. Le village immobile. Sennely-en-Sologne au XVIIIᵉ siècle. Paris, Plon. (*The immobile village. Sennely-en-Sologne in the 18ᵗʰ century*).

Bouman, P.J. – 1946. Geschiedenis van den Zeeuwschen Landbouw in de negentiende en twintigste eeuw en van de Zeeuwschen Landbouw-Maatschappij, 1843-1943, Wageningen, Veenman. (*History of agriculture in Zeeland in the 19ᵗʰ and 20ᵗʰ centuries*).

Bourke, A. – 1964. Emergence of potato blight, 1843-1846. Nature 203: 805-808.

Bourke, A. – 1993. 'The visitation of God'? The potato and the great Irish famine. Dublin, Lilliput Press.

Bourke, A., Lamb, H. – 1993. The spread of potato blight in Europe in 1845-6 and the accompanying wind and weather patterns. Dublin, Meteorological Service.

Bourson, Ph. – 1845. Exposé analytique de diverses opinions sur les causes probables de la maladie des pommes de terre, présenté à la commission instituée près du ministère de l'intérieur. Le Moniteur Belge 18: 2695-2699, 2703-2708, 2710-2715, 2719-2721. (*Analysis of various opinions on the probable causes of the potato disease ...*).

Braudel, F., Labrousse, E. (Eds.) – 1976. Histoire économique et sociale de France. Vol 3/1. L'avènement de l'ère industrielle (1789 – années 1880). Paris, Presses Universitaires de France. (*Economic and social history of France, Vol. 3-1. The coming of the industrial era (1789 – 1880s)*).

Braun, A. – 1846. Bemerkungen über den Spelzenrost des Roggens. Botanische Zeitung 4: 801-804. (*Observations on the glume rust of the rye*).

Brennan, L. – 1988. Government famine relief in Bengal, 1943. The Journal of Asian Studies 47: 541-566.

Broekhuizen, S. (Ed.) – 1969. Agro-ecological atlas of cereal growing in Europe. II. Atlas of the cereal-growing areas in Europe. Amsterdam, Elsevier. Wageningen, Pudoc.

Bruckmüller, E. – 2001. Sozialgeschichte Österreichs. 2nd Ed. Wien, Verlag für Geschichte und Politik. (*Social history of Austria*).

Brugger, H. – 1956. Die schweizerische Landwirtschaft in der ersten Hälfte des 19.Jahrhunderts. Frauenfeld, Huber. (*Swiss agriculture in the first half of the 19th century*).

Brugmans, H. – 1937. Geschiedenis van Nederland. Deel 6. Nieuwe Geschiedenis. Amsterdam, Joost Van den Vondel. (*History of the Netherlands. Part 6. New history*).

Brugmans, I.J. – 1929. De arbeidende klasse in Nederland in de 19e eeuw. 2nd ed. The Hague, Nijhoff. (*The labouring class in the Netherlands in the 19th century*).

Bulliard, J.B.F.P. – 1783. Dictionnaire élémentaire de botanique. Paris, Didot. (*Elementary dictionary of botany*).

Bulliard, J.B.F.P. – 1791. Herbier de la France. Seconde division. Histoire des champignons de la France. II. Paris, Didot. (*Herbarium of France. Second division. History of the fungi of France*).

Bumm, F. (Ed.) – 1928. Deutschlands Gesundheitsverhältnisse unter dem Einfluss des Weltkrieges. Stuttgart, Deutsche Verlags-Anstalt. Vol. I and II. (*Germany's health situation under the influence of the World War*).

Butler, E.J., Jones, S.G. – 1955. Plant pathology. Reprint. London, Macmillan.

Carefoot, G.L., Sprott, E.R. – 1967. Famine in the wind: man's battle against plant disease. Chicago, Rand McNally.

Chambry, É. – 1960. Ésope, Fables. Paris, Les Belles Lettres: fable nr. 58. (*Aesop, fables*).

Chaumartin, H. – 1946. Le mal des ardents et le feu Saint-Antoine. Vienne-la-Romaine, s.n. (*The itching disease and Saint Anthony's fire*).

Chester, K.S. – 1946. The nature and prevention of the cereal rusts as exemplified in the leaf rust of wheat. Waltham (Mass.), Chronica Botanica.

Clark, P.J. – 2007. Heliacal rising and setting of Sirius: 800 B.C. – 2000 A.D. http://www.angelfire.com/wizard/regulus_antares/heliacal_risings_and_settings_of.htm. consulted 30 January 2007.

Columella – see Forster & Heffner, 1968.

Committee – 1845. Verslag der Commissie van Landbouw in de provincie Groningen over de thans heerschende Aardappel-ziekte. Nederlandsche Staats-Courant N° 219, Tuesday 16 September, p. 2-3. (*Report of the Committee on Agriculture in the Province of Groningen on the presently prevailing potato disease*).

Connell, K.H. – 1950. The population of Ireland 1750-1845. Oxford, Clarendon Press.

Connell, K.H. – 1962. The potato in Ireland. Past and Present 23: 57-71.

Conquest, R. (Ed.) – 1984. The man-made famine in Ukraine. Washington, American Enterprise Institure.

Conquest, R. – 1986. The harvest of sorrow: Soviet collectivization and the Terror-Famine. London, Hutchinson.

Cummins, G.B. – 1971. The rust fungi of cereals, grasses and bamboos. Berlin, Springer.

Curtis, W. – 1805. Practical observations on the British grasses, especially as are best adapted to the laying down or improving meadows and pastures, 4th Ed. To which is now added a short account of the diseases in corn [= wheat], by Joseph Banks. London, Symonds.

Darwin, E. – 1800. Phytologia; or the philosophy of agriculture and gardening. With the theory of draining morasses, and with an improved construction of the drill plough. London, Johnson.

Davies, R.W., Wheatcroft, S.G. – 2004. The years of hunger: Soviet agriculture, 1931-1933. Basinstoke (Hamps.), Palgrave-Macmillan.

De Balzac, H. – 1844(?). Les paysans. Ed.1928. Paris, Flammarion. (*The peasants*).

De Bary, A. – 1853. Untersuchungen über die Brandpilze und die durch sie verursachten Krankheiten der Pflanzen mit Rücksicht auf das Getreide und andere Nutzpflanzen. Berlin, Müller. (*Investigations on the brand fungi and the diseases of plants caused by them with reference to cereals and other useful plants*, translated in 1969 as Phytopathological Classics 11).

De Bary, A. – 1861. Die gegenwärtig herrschende Kartoffelkrankheit, ihre Ursache und ihre Verhütung. Eine pflanzenphysiologische Untersuchung. Leipzig, Förstner. (*The presently prevailing potato disease, its cause and its prevention. A plant physiological study*).

De Bary, A. – 1866. Neue Untersuchungen über die Uredineen insbesondere die Entwicklung der *Puccinia graminis* und den Zusammenhang derselben mit *Aecidium berberidis*. Monatsbericht der Königlichen Preussischen Akademie der Wissenschaften (Session January 12th, 1865): 15-22. (*New studies on the Uredineae and especially the development of* Puccinia graminis *and the relation thereof with* Aecidium berberidis).

De Bary, A. – 1867. Neue Untersuchungen über Uredineen. Monatsbericht der Königlichen Preussischen Akademie der Wissenschaften (Session April 19th, 1866): 205-215. (*New studies on Uredineae*).

De Beus, J. – 2006. Resilient liberalism in a small Western European State, p. 75-98. In: P. Van Schie & G. Voerman (Eds.), The dividing line between success and failure. A comparison of liberalism in the Netherlands and Germany in the 19th and 20th centuries. Berlin, LIT Verlag.

De Bokx, J.A. – 1982. Bladrol en de bedreiging van de pootaardappelteelt door virussen. Gewasbescherming 13 (4/5): 23-32. (*Leaf curl and the threat by viruses to seed potato production*).

De Bosch Kemper, J. – 1851. Geschiedkundig onderzoek naar de armoede in ons vaderland, hare oorzaken en de middelen, die tot hare vermindering zouden kunnen worden aangewend.1e druk. Haarlem, Loosjes. (*Historical investigation into the poverty of our country, its causes and the means that could be used to its reduction*).

De Bosch Kemper, J. – 1882. Geschiedenis van Nederland na 1830. Vol. V. Amsterdam, Müller. (*History of the Netherlands after 1830*).

Decaisne, M.J. – 1846. Histoire de la maladie des pommes de terre en 1845. Paris, Dusacq. (*History of the potato disease in 1845*).

Decandolle, M. – 1807. Sur les champignons parasites. Annales du Muséum d'Histoire Naturelle 9: 56-74. *Ex* Seringe, 1818. (*On the parasitic fungi*).

De Candolle, M. – 1815. Flore Française. Paris, Desray. Tome cinquième, ou sixième volume. (*French flora*).

De Crescentio, P. – ~1300. De agricultura vulgare. Venice (1511), s.n. (*On common agriculture*).

Degreve, D. – 1982. Histoire quantitative et développement de la Belgique au XIXe siècle. V, 1a. Le commerce extérieur de la Belgique, 1830-1913-1939. Présentation critique des données statistiques. Bruxelles, Palais des Académies. (*Quantitative history and development of Belgium in the 19th century. External trade of Belgium 1830-1913-1939. Critical presentation of the statistical data*).

De Joode, T. – 1981. Landleven, het boerenbestaan van toen. Amsterdam, Elsevier. (*Country life, farmer existence in former days*).

Delacroix, G., Maublanc, A. – 1916. Maladies des plantes cultivées. Tome 2. Maladies parasitaires. Paris, Baillière. (*Diseases of cultivated plants. II. Parasitic diseases*).

De Meere, J.M.M. – 1982. Economische ontwikkeling en levensstandaard in Nederland gedurende de eerste helft van de negentiende eeuw. Aspecten en trends. Den Haag, Nijhoff. (*Economic development and standard of living in the Netherlands during the first half of the nineteenth century. Aspects and trends*).

Desmazières, J.B.H.J. – 1812. Agrostographie des départmens du Nord de la France. Lille, Vanackere. (*Agrostography of the departments of northern France*).

Desmazières, J.B.H.J. – 1847. Quatorzième notice sur les plantes cryptogames récemment découvertes en France. Annales des sciences naturelles, Série 3 Botanique Vol. 8: 9-37. (*Fourteenth note on the cryptogamous plants recently detected in France*).

Desmazières, J.B.H.J. – 1857. Vingt-quatrième notice sur les plantes cryptogames récemment découvertes en France. Bulletin de la Société Botanique de France 4: 797-803. (*Twenty-fourth note on the cryptogamous plants recently detected in France*).

Devoto, G. – 1948. Le tavole di Gubbio. Firenze, Sansoni. (*The tablets of Gubbio*).

Diamond, J. – 1997. Guns, germs, and steel. The fates of human societies. Norton, New York.

Diné, H. – 1951. La grande peur dans la généralité de Poitiers. Juillet-Août 1789. Paris, Diné. (*The great fear in the generality of Poitiers*).

Dirlmeier, U. – 1995. Kleine Deutsche Geschichte. Stuttgart, Reclam. (*Small German history*).

aDrèze, J., Sen, A. – 1989. Hunger and public action. Oxford, Clarendon Press.

Droz, J. – 1957. Les révolutions allemandes de 1848. Paris, Presses Uinversitaires de France. (*The German revolutions of 1848*).

Droz, J. – 1967. De la restauration à la révolution, 1815-1848. Paris, Armand Colin. (*From the restauration to the revolution, 1815-1848*).

Dumont, G.-H. – 1977. Histoire de la Belgique. Paris, Hachette. (*History of Belgium*).

Du Mortier, B.-C. (Dumortier) – 1845. Observations sur la cloque des pommes de terre. Bulletin de l'Académie Royale des Sciences et Belles-Lettres de Belgique 12: 285-299. (*Observations on the foliar shrivelling of the potatoes*).

Dupâquier, J. – 1980. Les aventures démografiques de la France et de l'Irlande (18e – 20e siècles), p. 167-180. In: L.M. Cullen & F. Furet (Eds.), Irlande et France XVIIe – XXe siècles. Pour une histoire rurale comparée. Paris, École des hautes études en sciences sociales. (*The demographic adventures of France and Ireland (18th to 20th centuries)*).

Dupâquier, J. (Ed.) – 1988. Histoire de la population française. Vol. 3. De 1789 à 1914. Paris, Presses Universitaires de France. (*History of the French populace*).

Dyson, T., Ó Gráda, C. – 2002. Famine demography. Perspectives from the past and present. Oxford, Oxford University Press.

Easton, C. – 1928. Les hivers dans l'Europe occidentale. Leiden, Brill. (*The winters in western Europe*).

Elsner, J.G. – 1847. Zusammenstellung des Ergebnisses der dieszjährigen Ernte und darauf begründete Muthmaszung für die Fruchtpreise. Oekonomische Neuigkeiten und Verhandlungen 74: 809-812. (*Compilation of the results of this year's harvest and the resulting expectation of the product prices*).

Enklaar, E.C. – 1846. De aardappelziekte van het jaar 1845. De Vriend van den Landman 10: 262-278. (*The potato disease of the year 1845*).

Enklaar, E.C. – 1847. Over de roest in het koorn. De vriend van de landman: 5324-635. (*On the rust in the corn [= rye]*).

Enklaar, E.C. – 1850. Handboek voor den akkerbouw, vrij bewerkt naar het Handbuch des Ackerbaues von William Löbe. Zwolle, Tjeenk Willink. (*Handbook for arable agriculture*).

Eriksson, J. – 1884. Om potatissjukan, dess historia och nature samt skyddsmedlen deremot. Stockholm, Norstedt. (*On the potato disease, its history and nature with the control methods thereof*).

Eriksson, J., Henning, E. – 1896. Die Getreideroste. Ihre Geschichte und Natur sowie Massregeln gegen Dieselben. Stockholm, Norstedt. (*The cereal rusts. Their history and nature and measures to control them*).

Eskens, E. – 2000. Filosofische reisgids voor Nederland en Vlaanderen. Amsterdam, Contact. (*Philosophical travel guide for the Netherlands and Flanders*).

Esmarch, F. – 1919. Zur Kenntnis des Stoffwechsels in blatrollkranken Kartoffeln. Zeitschrift für Pflanzenkrankheiten 29: 1-20. (*On the knowledge of the metabolism in potatoes diseased by leaf roll*).

Fabricius. J.C. – 1774. Attempt at a dissertation on the diseases of plants. Phytopathological Classics 1 (1926): 1-66. (Translated by M. Kølpin Ravn from the Danish original).

Festus – see Lindsay, W.M. (1965).

Finkelstein, I., Silberman, N.A. – 2002. The Bible unearthed. Archaeology's new vision of ancient Israel and the origin of its sacred texts. New York, Simon & Schuster.

Fischer, J. – 1915. Wie sorgen wir für Einbringung der Ernte 1915? Anregungen zur Lösung einer Lebensfrage des deutschen Volkes. Fühling's Landwirtschaftliche Zeitung 64: 315-321. (*How can we secure to get in the 1915 harvest? Suggestions to solve a vital question of the German people*).

Flemming, J. – 1978. Landwirtschaftliche Interessen und Demokratie. Bonn, Neue Geselschaft. (*Agricultural interests and democracy*).

Fontana, F. – 1767. Observations on the rust of grain. Phytopathological Classics 2 (1932): 1-40.

Forster, E.S., Heffner, E.H. – 1968. Lucius Junius Moderatus Columella. On agriculture and trees. London, Heinemann.

Foucault, M. – 1966. Les mots et les choses. Paris, Gallimard. (*The words and the things*).

Frazer, J.G. – 1989. Ovid V. Fasti, 2nd Ed. Cambridge (Mass.), Harvard University Press.

Frey, M. – 1998. Der Erste Weltkrieg und die Niederlande. Ein neutrales Land im politischen und wirtschaftlichen Kalkül der Kriegsgegner. Berlin, Akademie Verlag. (*The First World War and the Netherlands. A neutral country in the political and economic calculation of warring opponents*).

Fries, E. – 1829. Systema mycologicum, sistens fungorum ordines, genera et species. Vol III. Et ultimum. Greifswald, Mauritius. (*Mycological system, consisting of orders, genera and species*).

Fruwirth, C. – 1916. Ein Fall von Taubährigkeit. Wiener landwirtschaftliche Zeitung 66: 365. (*A case of deaf ears*).

Fuchs, J.M. – 1970. Amsterdam een lastige stad. Rellen, oproeren en opstanden in de loop der eeuwen. Baarn, de Boekerij (*Amsterdam, a naughty town; disturbances, riots and insurgencies in the course of the centuries*).

Fuckel, L. – 1869. Symbolae mycologicae. Beiträge zur Kenntniss der Rheinischen Pilze. Wiesbaden, Niedner. (*Mycological symbols. Contributions to the knowledge of the fungi in Rhineland*).

Furet, F. – 1988. La révolution, de Turgot à Jules Ferry, 1770-1880. Paris, Hachette. (*The revolution, from Turgot to Jules Ferry, 1770-1880*).

Furet, F., Ozouf, M. – 1988. Dictionnaire critique de la révolution française. Paris, Flammarion. (*Critical dictionary of the french revolution*).

Furet, F., Richet, D. – 1973. La révolution française. Paris, Fayard. (*The French revolution*).

Fürnrohr, XX. – 1845. Über die Krankheit der Kartoffeln. Oekonomische Neuigkeiten und Verhandlungen, Prag 69: 934-935. (*On the disease of the potatoes*)

Gadichaud, Ch. – 1846. Aperçu sur les causes physiologiques de la maladie des pommes de terre. Comptes Rendus hebdomadaires des Séances de l'Académie des Sciences 22: 271-281. (*Summary of the physiological causes of the potato disease*).

Galli, E. – 1924. Castiglioncello, scoperte di antichità varie, compresa un'ara referibile al culto di Robigus. Notizie degli Scavi: 157-178. (*Castiglioncello, discovery of various antiquities, including an altar referring to the cult of Robigo*).

Gäumann, E. – 1946. Pflanzliche Infektionslehre. Basel, Birkhäuser. (*Principles of plant infection*).

Geertsema, C.J. – 1868. Beschrijving van den landbouw in de districten Oldambt, Westerwolde en Fivelgo in de provincie Groningen. Tijdschrift ter Bevordering van Nijverheid 31 (3e reeks #9): 49-112/131-194/211-292. (*Description of the agriculture in the districts Oldambt, Westerwolde and Fivelgo in the province Groningen*).

General Report – see 'Algemeen Verslag'.

Gianferrari, A. – 1995. Robigalia: Un appuntamento per la salvezza del raccolto, p. 127-140. In: Agricoltura e commerci nell'Italia antica / [a cura di: Lorenzo Quilici e Stefania Quilici Gigli]. (Atlante tematico di topografia antica Supplemento 1). Roma. (*Robigalia, an agreement to save the harvest*).

Gieysztor, A., Kieniewicz, S., Rostworonski, E., Tazbir, J., Wereszycki, H. – 1979. History of Poland. Warszawa, Polish Scientific Publishers.

Goebel, G. – 1980. Textanalytischer Versuch zu einem Abschnitt aus Balzacs Les Paysans, p. 437-460. In: Gumbrecht, H-U., Stierle, K. & Warning, R. (Eds.), Honoré de Balzac. München, Fink. (*Experiment in text analysis of a fragment from Balzac's The Peasants*).

Gómez-Alpizar, L., Carbone, I., Ristaino, J.B. – 2007. An Andean origin of *Phytophthora infestans* inferred from mitochondrial and nuclear gene genealogies. PNAS 104: 3306-3311.

Ghose, R.L.M., Ghatge, M.B., Subrahmanyan, V. – 1960. Rice in India. New Delhi, ICAR.

Goossen, Th. – 1980. Hagelkruisen, voorwerpen met religieuze en profane betekenis. Nederlands Volksleven 30 (3/4): 64-75. (*Hail crosses, objects with religious and profane meanings*).

Göppert, X.X. – 1846. Der Spelzenbrand im Roggen. Botanische Zeitung 4: 545-548. (*The glume blight in rye*).

Goswami, O. – 1990. The Bengal Famine of 1943: re-examining the data. Indian Economic and Social History Review 27 No 4.

Gotthelf, J. [= Albert Bitzius] – 1847. Käthi die Grossmutter. Reprint 1965. Erlenbach/Zürich, Rentsch. (*Cathy, the grandmother*).

Goubert, J-P. – 1974. Malades et médecins en Bretagne 1770-1790. Rennes, Institut de Recherches historiques. (*Patients and medical practitioners in Brittany, 1770-1790*).

Govers, F, Gijzen, M. – 2006. *Phytophthora* genomics: the Plant Destroyers' genome decoded. Molecular Plant-Microbe Interactions 19: 1295-1301.

Graham, S. -1929. Peter the Great. A life of Peter I of Russia called the great. London, Ernest Benn.

Gram, E. (Ed.) – 1940. Kartoflens Sygdomme i Billeder og Tekst. København, Reitzels. (*Potato diseases in pictures and text*).

Greenough, P.R. – 1982. Prosperity and misery in modern Bengal. The famine of 1943-1944. New York, Oxford University Press.

Grootegoed, H. – 1853. Bijdragen tot de kennis van den akker- en tuinbouw van het Westland in Zuid-Holland. Tijdschrift ter Bevordering van Nijverheid 16 (2e reeks #1) 24-39. (*Contributions to the knowledge of agriculture and horticulture of the Westland in Zuid Holland*).

Gruner, E. – 1968. Die Arbeiter in der Schweiz im 19. Jahrhundert. Soziale Lage, Organisation, Verhältnis zu Arbeitgeber und Staat. Bern, Francke. (*The labourers in Switzerland in the 19th century ...*).

Guin, Y. – 1982. Histoire de la Bretagne de 1789 à nos jours. Paris, Maspero. (*History of Brittany from 1789 to our days*).

Harting, P. – 1846. Recherches sur la nature et les causes de la maladie des pommes de terre en 1845. Verhandelingen der 1e Klasse van het Koninklijk Nederlandsch Instituut van Wetenschappen etc., Amsterdam. Nieuwe Verhandelingen 12: 203-297 met 3 gekleurde platen. (*Studies on the nature and causes of the potato disease*). Also published as: Harting, P. – 1846. Extrait d'un mémoire sur la nature et les causes de la maladie des pommes de terre en 1845. Annales des Sciences Naturelles, 3me série. Botanique. 6: 42-62.

Heesterbeek, J.A.P., Zadoks, J.C. – 1987. Modelling pandemics of quarantine pests and diseases; problem and perspectives. Crop Protection 6: 211-221.

Heldring, O.G. – 1845. De nood en hulp der armen, in betrekking tot den arbeid, de weelde en het medelijden. Eenige praktische blikken in den toestand onzes Volks. Amsterdam, Beijerinck. (*Needs and relief of the poor, in relation to labour, affluence and pity. Some practical views on the condition of our people*).

Heldring, O.G. – 1847. Opmerkingen op eene reis langs den Rijn. Amsterdam, Beijerinck. (*Comments on a trip along the Rhine*).

Hermannsen, J.E. – 1968. Studies on the spread and survival of cereal rust and mildew diseases in Denmark. Contributions of the Department of Plant Pathology, Royal Veterinary and Agricultural College, Copenhagen 87: 1-206.

Herzfeld, H. – 1968. Der Erste Weltkrieg. München, Deutscher Taschenbuch Verlag. (*The First World War*).

Hilmer, F. – 1916. Kriegswirtschaftliche Massnahmen in Oesterreich. Mitteilungen der deutschen Landwirtschafts-Gesellschaft 32: 822-823. (*War-economic measures in Austria*).

Hiltner, L. – 1905. Über die Getreideroste, unter besonderer Berücksichtigung ihres Auftretens im Jahre 1904. Praktische Blätter Pflanzenbau N.F. 3: 39-43; 54-55; 64-66; 79-82. (*On the cereal rusts, with special attention for their occurrence in the year 1904*).

Hirschfeld, G., Krumeich, G., Renz, I. – 2003. Enzyklopädie Erster Weltkrieg. Paderborn, Schöningh. (*Encyclopedia of the First World War*).

Hlubek, F.X. – 1847. Die Kartoffelkrankheit in Steiermark, Kärnthen und Krain im Jahre 1846. Oekonomische Neuigkeiten und Verhandlungen 73: 65-71. (*The potato disease in Styria, Carinthia and Carniola [Kranjska] in the year 1846*).

Hoekstra, M.E. – 1879. Uit het dagboek van Mindert Everts Hoekstra (1825-1879). (*From the diary of M.E. Hoekstra (1825-1979)*). http://irnsum.tmfweb.nl. Consulted 041205.

Hofstee, E.W. – 1978. De demografische ontwikkeling van Nederland in de eerste helft van de negentiende eeuw. Een historisch-demografische en sociologische studie. Deventer, Van Loghum Slaterus. (*The demographic development of the Netherlands in the first half of the nineteenth century* ...).

Hogg, W.H., Hounam, C.E., Mallik, A.K., Zadoks, J.C. – 1969. Meteorological factors affecting the epidemiology of wheat rusts. World Meteorological Organization, Techn. Note No. 99, Geneva, Switzerland.

Holdefleisz, P. – 1917. Eine weitere Beitrag über die Ursachen der Kartoffelmissernte 1916. Illustrierte Landwirtschaftliche Zeitung 37: #15 dd 21-02-1917. (*A further contribution on the causes of the potato crop failure 1916*).

Honcamp, F. – 1918. Die deutsche Landwirtschaft und die Ernährung des deutschen Volkes. Fühlings Landwirtschaftliche Zeitung 433-442. (*The German agriculture and the food supply of the German people*).

Hoogerhuis, O.W. – 2003. Baren op Beveland. Vruchtbaarheid en zuigelingensterfte in Goes en omliggende dorpen gedurende de 19e eeuw. Wageningen, A.A.G. Bijdragen 42. (*Giving birth on Beveland. Fertility and infant mortality in Goes and surrounding villages during the 19th century*).

Hooijer, C. – 1847. De groote nood des hongers in en bij den Boemelerwaard. 2nd ed. Zalt-Bommel, Noman. (*The great distress of the hunger in and near the Bommelerwaard*)

Hooper, W.D., Boyd Ash, H. – 1967. Marcus Terentius Varro on agriculture. London, Heinemann.

Horsfall, J.G., Cowling, E.B. – 1978. Some epidemics man has known, p. 17-32. In: Horsfall, J.G., Cowling, E.B. (Eds.), Plant disease, an advanced treatise. Volume III. How disease develops in populations. New York, Academic Press

Hort, A. – 1916. Theophrastus, enquiry into plants, I and II. London, Heinemann.

Houssel, J.-P. (Ed.) – 1976. Histoire des paysans français du XVIIIe siècle à nos jours. Roanne, Horvath. (*History of French peasants from the 18th century up to our days*).

Huijboom, H. – 1992. Het aardappeloproer van 1917. Ons Amsterdam 44: 148-152. (*The potato revolt of 1917*).

Hylander, N., Jørstad, I, Nannfeldt, J.A. – 1953. Enumeratio uredinearum scandinavicarum. Opera botanica 1: 1-102. (*A listing of Scandinavian rusts*).

Ising, A.L.H. – 1892. Verslag der Handelingen van de Staten-Generaal gedurende de zitting van 1845-1846. Den Haag, Nijhoff. (*Parliamentary Reports of the session 1845-1846*).

Jansen, P.C., de Meere, J.M.M. – 1982. Het sterftepatroon in Amsterdam, 1774-1930, een analyse van de doodsoorzaken. Tijdschrift voor sociale geschiedenis 8: 180-223. (*The mortality pattern in Amsterdam, 1774-1930, an analysis of the causes of death*).

Jansma, K., Schroor, M. (Eds) – 1987. Tweehonderd jaar geschiedenis van de Nederlandse Landbouw. Leeuwarden, Inter-Combi van Seyen. (*Two hundred years of history of Dutch agriculture*).

Jardin, A., Tudesq, A.J. – 1973. La France des notables. L'évolution générale 1815-1848. Paris, Éd. du Seuil. (*The France of the dignitaries; The general evolution 1815-1848*).

Kangle, R.P. – 1986. The Kauṭilīya arthaśāstra, Part II, 1.19.34. Bombay, University of Bombay (1963), reprint Delhi (1968).

Kaplan, S.L. – 1982. The famine plot persuasion in eighteenth-century France. Transactions of the American Philosophical Society 72 (3): 1-79.

Karsten, P.A. – 1879. Mycologia fennica. IV. Hypodermii, Phycomycetes et Myxomycetes. Bidrag till kännedom af Finlands natur och folk 31: 1-143. (*Finnish mycology ...*).

Kent, R.G. – 1967. Varro on the Latin language. I. Books V.-VII. London, Heinemann.

Kickx, J. – 1867. Flore cryptogamique des Flandres. Vol. 2. Gand, Hoste. (*Cryptogamic flora of Flanders*).

Kielmansegg, P. Graf – 1980. Deutschland und der Erste Weltkrieg. 2nd Ed. Stuttgart, Klett-Cotta. (*Germany and the first World War*).

Kislev, M.E. – 1982. Stem rust of wheat 3300 years old found in Israel. Science 216: 993-994.

Klomp, A.O. – 1977. Early senescence of rice and *Drechslera oryzae* in the Wageningen Polder, Surinam. Wageningen, Pudoc.

Koeman, J.H., Zadoks, J.C. – 1999. History and future of plant protection policy, from ancient times to WTO-SPS, p. 21-48. In: G. Meester, R.D. Woittiez, A. de Zeeuw (Eds.), Plants and politics. Wageningen, Wageningen Pers.

Kølpin Ravn, F. – 1914. Smitsomme sygdomme hos Landbrugsplanterne. Copenhagen, August Bang. (*Infectious diseases of arable crops*).

Kops, J. – 1808. No title. Magazijn van Vaderlandsche Landbouw 5: 73-149. Haarlem, Loosjes.

Kops, J. – 1810. No title. Magazijn van vaderlandschen Landbouw. V. Deel. Haarlem, Loosjes.

Kornauth, K. – 1915. Mitteilungen der Wiener Pflanzenschutzstation. Sonder-Zeitschrift für das Landwirtschaftlichen Versuchswesen in Österreich, Berichte der k.k. landwirtschaftlich-bakteriologischen und Pflanzenschutsstation im Wien, 1914. Abstract in Zeitschrift für Pflanzenkrankheiten 25: 463-464. (*Communications of the Viennese Institute of Plant Protection*).

Kornauth, K. – 1917. Bericht über die Tätigkeit der k.k. landwirtschaftlichen-bakteriologischen und Pflanzenschutzstation in Wien im Jahre 1916. Zeitschrift für das Landwirtschaftliche Versuchswesen in Österreich 20: 288-314. Abstract in Zeitschrift für Pflanzenkrankheiten 28 (1918): 213-214. (*Report on the activities of the imperial-royal agricultural-bacteriological and plant protection station in Vienna in the year 1916*).

Kornauth, K. – 1918. Bericht der k.k. landwirtschaftlichen-bakteriologischen und Pflanzenschutzstation in Wien für das Jahr 1917. Zeitschrift für das Landwirtschaftliche Versuchswesen in Österreich 21: 377-393. Abstract in Zeitschrift für Pflanzenkrankheiten 29 (1919): 242-243. (*Report on the activities of the imperial-royal agricultural-bacteriological and plant protection station in Vienna in the year 1917*).

Kornauth, K. – 1919. Bericht der k.k. landwirtschaftlichen-bakteriologischen und Pflanzenschutzstation in Wien für das Jahr 1918. Zeitschrift für das landwirtschaftliche Versuchswesen in Österreich 22: 28-44. Abstract in Zeitschrift für Pflanzenkrankheiten 30 (1920): 83-84. (*Report on the activities of the imperial-royal agricultural-bacteriological and plant protection station in Vienna in the year 1918*).

Kossman, E.H., Krul, W.E. – 1977. Winkler Prins Geschiedenis der Nederlanden, Vol. 3: 1780-1870. Amsterdam, Elsevier. (*Winkler Prins History of the Netherlands ...*).

Kotansky, R. – 1994. Magic in Roman North Africa. Available at: http://ccat.sas.upenn.edu/jod/apuleius/renberg/MAINTEXT.HTML. Accessed August 2008.

Kraus, A. – 1980. Quellen zur Bevölkerungs-, Sozial- und Wirtschaftsstatistik Deutschlands 1815-1875. I. Boppard am Rhein, Harald Boldt. (*Sources for the demographic, social and economical statistics of Germany, 1815-1875*).

Kroes, P. – 1987. Sociale onrust in Zeeland 1845-1846. Middelburg, Cie Regionale Geschiedbeoefening. (*Social unrest in Zeeland 1845-1846*). Report with limited circulation.

Kruizinga, J. (Ed.) – 2002. Het XYZ van Amsterdam. Amsterdam, Amsterdam Publishers. (*The XYZ of Amsterdam*).

Kühn, J. – 1858. Die Krankheiten der Kulturgewächse, ihre Ursachen und ihre Verhütung. Berlin, Bosselmann. (*The diseases of the agricultural crops, their causes and their prevention*).

Kuhn, J. – 1918. Die Gefahr einer neuen Kartoffelmiszernte. Deutsche Landwirtschaftliche Presse 45: 196, 200-201. (*The danger of another potato crop failure*).

References

Kyrre, H. – 1913. Kartoffelens krønike, en kulturhistorisk studie. København, Gad. (*Potato chronicle, a study in cultural history*).

Lamb, H.H. – 1995. Climate, history and the modern world. 2nd Ed. London, Routledge.

Lambert, A.B. – 1798. Description of the blight of wheat, *Uredo frumenti*. Transactions of the Linnean Society (London) 4: 193-194.

Lamberty, M. – 1949. Geschiedenis van Vlaanderen, Vol. VI. Amsterdam, Joost Van den Vondel. (*History of Flanders*).

Lamberty, M., Lissens, R.F. – 1951/2. Vlaanderen door de eeuwen heen. Brussels, Elsevier. Vol. I - 1951, Vol. II – 1952. (*Flanders throughout the centuries*).

Large, E.C. – 1940. The advance of the fungi. London, Jonathan Cape.

Le Couteur, P., Burreson, J. – 2003. Napoleon's buttons: how 17 molecules changed history. New York, Tarcher.

Lefèbvre, G. – 1932. La grande peur de 1789. (Reprint 1957) Paris, Société d'Édtion d'Enseignement Supérieur. (*The great fear of 1789*).

Legge, J. – 1970. The works of Mencius. New York, Dover. (Replication of the first edition, 1895. Oxford, Clarendon Press.)

Lehmann, E., Kummer, H., Dannenmann, H. – 1937. Der Schwarzrost, seine Geschichte, seine Biologie und seine Bekämpfung in Verbindung mit der Berberitzenfrage. München, Lehmann. (*The black rust, its history, biology and control in relation to the barberry question*).

Leonhard, J. – 2006. Co-existence and conflict: structures and positions of nineteenth-century liberalism in Germany, p. 9-34. In: P. Van Schie & G. Voerman (Eds.), The dividing line between success and failure. A comparison of liberalism in the Netherlands and Germany in the 19th and 20th centuries. Berlin, LIT Verlag.

Leopold, H.M.R. – 1926. Het moederkoren en de rode hond, p. 263-268. In: 'Uit de leerschool der spade' Vol. III. Zutphen, Thieme. (*The ergot and the red dog*).

Le Roy Ladurie, E. – 1967. Histoire du climat depuis l'an mil. Paris, Flammarion. (*History of the climate since the year thousand*).

Le Roy Ladurie, E. – 2004. Histoire humaine et comparée du climat. I. Canicules et glaciers XIIIe – XVIIIe siècle). Fayard, Paris. (*Human and comparative history of the climate. I. Dog days and glaciers 13th to 18th century*).

Lestiboudois, T. – 1845. Rapport sur la maladie des pommes de terre. Mémoires de la Société Royale des Sciences, de l'Agriculture et des Arts de Lille 24 (1845) 245-278. (*Report on the potato disease*).

Lette, XX. – 1847. Auszug aus einem an des Herrn Ministers des Innern Excellenz erstatteten Reiseberichte des Präsidenten Lette über die Bereisung der Provinz Preuszen. Annalen der Landwirthschaft 5 (Vol. 10): 1-57. (*Abstract from a report ... by President Lette on travelling the Province Prussia*).

Léveillé, J.-H. – 1848. Urédinées, p. 768-790. In: C. d'Orbigny (Ed.), Dictionnaire universel d'histoire naturelle. Tome 12. Paris, Renard. (*The uridineae*).

Lind, J. – 1913. Danish fungi as represented in the herbarium of E. Rostrup. Copenhagen, Nordisk Verlag.

Lind, J. – 1916a. Kartoflernes Bladrullesyge. Ugeskrift for Landmænd 61: 62. (*Leaf-roll disease of potato*).

Lind, J. – 1916b. Forsøg med Anvendelse af Sprøjtemidler mod Kartoffelskimmel i Aarene 1910-1915. Tidskrift for Planteavl 23: 365-397. Abstract in Zeitschrift für Pflanzenkrankheiten 27: 157. (*Experiment on the use of spray products against the potato fungus in the years 1910-1915*).

Lindemans, P. – 1952. Geschiedenis van de landbouw in België. II. Antwerpen, de Sikkel. (*The history of agriculture in Belgium*).

Lindsay, W.M. – 1965. Sextus Pompeius Festus. De verborum significatu quae supersunt cum Pauli epitome. Bibliotheca Teubneriana. Hildesheim, Georg Olms. Reprint from 1913 edition). (*On the meaning of words ...*).

Linnaeus, C. – 1751. Philosophia botanica in qua explicantur fundamenta botanica cum definitionibus partium, exemplis terminorum, observationibus rariorum, adjectis figures aeneis. Stockholm, Kiesewetter. (*Botanical philosophy in which the botanical basics are explained, with definitions of parts, examples of terms …*).

Löbe, W. – 1847. Ernteberichte vom Jahre 1846. Oekonomische Neuigkeiten und Verhandlungen 73: 121-126, 131-135. (*Harvest reports from the year 1846*).

Löhr vom Wachendorf, F. - 1954. Die grosse Plage. Frankfurt a.M., Herkul. (*The great pests*).

Löw, I. – 1928. Die Flora der Juden. I. Kryptogamae, Acanthaceae, Graminaceae. Wien, Löwit. (*The flora of the Jews …*).

Lutz, H. – 1985. Die Deutschen und ihre Nation. 2. Zwischen Habsburg und Preussen. Deutschland 1815-1866. Berlin, Siedler. (*The Germans and their nation. 2. Between Habsburg and Prussia. Germany 1815-1866*).

Macartney, G.A. – 1968. The Habsburg Empire 1790-1918. London, Weidenfeld & Nicolson.

Mangin, L. – 1914. Parasites végétaux des plantes cultivées. Paris, Maison Rustique. (*Vegetable parasites of the cultivated plants*).

Mansergh, N. (Ed.) – 1973. Constitutional relations between Britain and India. The transfer of power 1942-7. Vol. IV. The Bengal Famine and the new Viceroyalty, 15 June 1943 – 31 August 1944. London, His Majesty's Stationary Office.

Marschalck, P. – 1973. Deutsche Überseewanderung im 19 Jahrhundert. Ein Beitrag zur soziologische Theorie der Bevölkerung. Stuttgart, Klett. (*German overseas migration in the 19th century …*).

Marshall, W. – 1795. The rural economy of Norfolk; comprising the Management of Landed Estates, and the present practice of husbandry in that county. Vol. II. 2nd Ed. London, Nicole.

Martens, XX. – 1845. Sur la maladie des pommes de terre. Bulletin de l'Académie royale des sciences, des lettres et des beaux-arts de Belgique 12: 356-372. (*On the disease of the potato*).

Massenot, M. – 1961. Épidémiologie de la rouille noire des céréales en France. CR hebd. Séances de l'Académie d'Agriculture de France 47: 54-600. (*Epidemiology of black rust on cereals in France*).

Matossian, M.K. – 1989. Poisons of the past. Molds, epidemics, and history. New Haven, Yale Uinversity Press.

Matsuo, T., Kumazawa, K., Ishii, R., Ishihara, K., Hirata, H. (Eds) – 1995. Science of the rice plant. II. Physiology. Tokyo, Food and Agriculture Policy Research Center.

Mauz, E.F. – 1845. Versuche und Beobachtungen über den Kartoffelbau und die Krankheiten der Kartoffeln, besonders im Jahr 1845. Stuttgart, Steinkopf. (*Experiments and observations on potato cultivation and potato diseases, especially in the year 1845*).

Mayer, A. – 1882a. Over de mozaïekziekte van de tabak; voorlopige mededeeling. Tijdschrift Landbouwkunde (Groningen) 2: 359-364. (*On the mosaic disease of tobacco; preliminary communication*).

Mayer, A. – 1882b. Über die Mosaikkrankheit des Tabaks. Landwirtschaftliche Versuchsanstalten 32: 451-467. (*On the mosaic disease of the tobacco*).

Meher, D. – 1917. Brotgetreideernte und Brotgetreideverbrauch in Friedens- und Kriegszeit. Statistisches Jahrbuch für das Deutsche Reich. (*Harvest of bread grain and use of bread grain in peace and war time*).

Mentzel, XX. – 1848. Bericht des Wirklichen Geheimen Kriegs-Rath Mentzel über eine nach den Oesterreichischen Staaten im Herbst 1846 unternommen Privatreise. Annalen der Landwirthschaft 6 (Vol. 11): 75-142. (*Report by … Mentzel on a private trip to the Austrian States in the autumn of 1846*).

Meyen, F.J.F. – 1841. Pflanzen-Pathologie. Lehre von dem kranken Leben und Bilden der Pflanzen. Berlin, Haude & Spener. (*Plant pathology. Knowledge of the diseased life and strife of plants*).

Mickle, J. – 1845. In: Home correspondence. The Gardeners' Chronicle: 592-593.

References

Mohorst, G. – 1977. Wirtschaftswachstum und Bevölkerungsentwicklung in Preussen 1816 bis 1914. Zur Frage demo-ökonomischer Entwicklungszusammenhänge. New York, Arno Press. (*Economic growth and demographic development in Prussia 1816 to 1914 ...*).

Moleschott, J., von Baumhauer, E.H. – 1845. Het wezen der aardappelziekte en de middelen ter voorkoming en genezing van dezelve. Utrecht, Bötticher. (*The nature of the potato disease and the means to prevent and to cure it*).

Mommsen, W.J. – 1984. Deutsche Geschichte 10: Bismarckreich und Wilhelminische Zeit. 1871-1918. Gütersloh, Berttelsmann. (*German history 10: the reign of Bismarck and the Williams' period*).

Mommsen, W.J. – 1995. Bürgerstolz und Weltmachtstreben. Deutschland unter Wilhelm II. 1890-1918. Berlin, Propyläenverlag. (*Civilian pride and the strife for world power. Germany under William II. 1890-1918*).

Mommsen, W.J. – 1998. Die ungewollte Revolution. Die revolutionäre Bewegungen in Europa 1830-1849. Frankfurt am Main, Fischer. (*The unintentional revolution. The revolutionary movements in Europe 1830-1849*).

Mommsen, W.J. – 2004. Die Urkatastrophe Deutschlands. Der Erste Weltkrieg 1914-1918. Vol. 17 of Gebhardt: Handbuch der Deutsche Geschichte. Stuttgart, Klett-Cotta. (*The primal catastrophe of Germany. The First World War 1914-1918*).

Morier Evans, D. – 1848. The commercial crisis 1847-1848; being facts and figures illustrative of the events of that important period, considered in relation to the three epochs of the railway mania, the food and money panic, and the French revolution. London, Letts, Son & Steer.

Montagne, C. – 1845. Observations sur la maladie des pommes de terre. L'institut, Journal universel des Sciences, Paris N° 609: 312-313. (*Observations on the disease of the potato*).

Morren, Ch. – 1845a. Instructions populaires sur les pommes de terre. Instructions populaires sur les moyens de combattre et de détruire la maladie actuelle (gangrène humide) des pommes de terre et sur les moyens d'obtenir pendant l'hiver des récoltes de ces tubercules, suivies de renseignements sur la culture et l'usage du topinambour. Bruxelles, Périchon. (*Popular instructions on the potato ...*).

Morren, Ch. – 1845b. [Comment following the note by Martens, 1845]. Bulletin de l'Académie royale des sciences, des lettres et des beaux-arts de Belgique 12 (Part 2): 372-373.

Morren, Ch. – 1845c. Notice sur le Botrytis dévastateur ou le champignon des pommes de terre. Annales de la Société royale d'Agriculture et de Botanique de Gand 1: 287-292. (*Note on the destructive Botrytis or the potato fungus*).

Morstatt, H. – 1925. Entartung, Altersschwäche und Abbau bei Kulturpflanzen, insbesondere der Kartoffel. Freising-München, Datterer. *Degeneration, old-age-weakness and running-out in cultivated plants, in the potato especially*).

Müller, H.C., Molz, E. - 1917. Über das Auftreten des Gelbrostes (*Puccinia glumarum*) am Weizen in den Jahren 1914 und 1916. Fühlings landwirtschaftliche Zeitung 66: 42-55. (*On the appearance of yellow rust (*Puccinia glumarum*) on wheat in the years 1914 and 1916*).

Münter, J.– 1846. Die Krankheiten der Kartoffeln insbesondere die im Jahre 1845 pandemisch herrschenden nasse Fäule. Berlin, Hirschwald. (*The diseases of potato particularly the wet rot pandemically prevailing in the year 1845*).

Neger, F.E. – 1919. Die Blattrollkrankheit der Kartoffel. Zeitschrift für Pflanzenkrankheiten 29: 27-48. (*The leaf roll disease of the potato*).

Newman, E.L., Simpson, R.K. (Eds.) – 1987. Historical dictionary of France from the 1815 Restoration to the Second Empire. Westport (Conn.), Greenwood.

Nitsch, J. – 1846. Landwirthschaftlicher Bericht aus Österr.-Schlesien. Oekonomische Neuigkeiten und Verhandlungen 70: 289-294. (*Agricultural report from Austrian Silesia*).

Oberstein, O. – 1914. Beobachtungen und Bekämpfung von Pflanzenkrankheiten in Schlezien. Abstract in Zeitschrift für Pflanzenkrankheiten 25 (1915) 400. (*Observations and control of plant diseases in Silesia*).

Ó-Gráda, C. -1989. The great Irish famine. London, Macmillan.

Oliemans, W.H. – 1988. Het brood van de armen. De geschiedenis van de aardappel temidden van ketters, kloosterlingen en kerkvorsten. Den Haag, SDU. (*The bread of the poor. The history of the potato amidst heretics, monks, and princes of the church*).

Oort, A.J.P. – 1941. De berberis een gevaar voor de graancultuur? Tijdschrift over Plantenziekten (European Journal of Plant Pathology) 47: 112-119. (*The barberry a danger for the grain cultivation?*).

Oortwijn Botjes, J.G. – 1920. De bladrolziekte van de aardappel. Wageningen, Veenman. Ph.D. Thesis. (*The leaf roll disease of the potato*).

Ordish, G. – 1976. The constant pest. A short history of pests and their control. London, Peter Davis.

Orlob, 1973. Frühe und mittelalterliche Pflanzenpathologie. Pflanzenschutz Nachrichten Bayer 26: 69-314. (*Early and medieval plant pathology*).

Ou, S.H. – 1985. Rice diseases. 2nd Ed. Kew, Commonwealth Agricultural Bureaux.

Ovid – *Fasti*, see Frazer, J.G. (1989).

Padmanabhan, S.Y. – 1973. The Great Bengal Famine. Annual Review of Phytopathology 11: 11-26.

Padmanabhan, S.Y., Roy Chowdhry, K.R., Ganguly, D. – 1948. *Helminthosporium* disease of rice. I. Nature and extent of damage caused by the disease. Indian Phytopathology 1: 34-47.

Palou, J. – 1955. La Grande Peur de 1789 en Oisans. Annales Historiques de la Révolution Française 27: 50-54. (*The Great Fear of 1789 in Oisans*).

Parmentier, A. – 1781. Recherches sur les végétaux nourrissans, qui, dans les temps de disette, peuvent remplacer les alimens ordinaires. Paris, de Prony. (*Studies on nutricious plants that, in times of scarcity, can replace the usual foods*).

Persoon, D.C.H. – 1801. Synopsis methodica fvngorvm. Göttingen, Dieterich. (*Methodical overview of the fungi*).

Peterson, P.D. – 1995. The influence of the potato blight epidemics of the 1840s on disease etiology theory in plants, p. 30-35. In: Dowley, L.J., Bannon, E., Cooke, L.R., Keane, T., O'Sullivan, E. (Eds.), *Phytophthora infestans* 150. Dublin, Boole.

Pinckert, F.A. – 1867. Der Roggen und Dinkel. Stuttgart, Johannsen. (*Rye and spelt*).

Pirenne, H. – 1928. La Belgique et la Guerre Mondiale. Paris. Presses Universitaires de France. (*Belgium and the World War*).

Pirenne, H. – 1932. Histoire de la Belgique. VII. De la révolution de 1830 à la guerre de 1914. Bruxelles, Lamartin. (*History of Belgium. VII. From the revolution of 1830 to the war of 1914*).

Plenck, J.J. – 1794. Physiologia et pathologia plantarum. Vienna, Bluauer. (*Physiology and pathology of plants*).

Pliny *NH*– see Rackham, H. – 1971. (*NH = Naturalis Historia*).

Pogge, F. - 1893. Zur Rostfrage. Zeitschrift für Pflanzenkrankheiten 3: 57-60. (*On the rust question*).

Ponse, H. – 1810. Leerboek over den landbouw, in zamenspraken. 1. Leyden, du Mortier. (*Textbook on the agriculture, in conversations*).

Ponse, H. – 1827. Verhandeling over den honigdaauw. Middelburg, Van Benthem. (*Treatise on the honeydew*).

Postan, M.M. (Ed.) – 1966. The Cambridge economic history of Europe. I. The agrarian life of the Middle Ages. Cambridge, Cambridge University Press.

Powell, J. (Ed.) – 1998. Chronology of European History 15,000 B.C. to 1997. Vol. 2. 1765-1997. Chicago, Fitzroy Dearborn.

Priester, P. – 1991. De economische ontwikkeling van de landbouw in Groningen 1800-1910. Een kwalitatieve en kwantitatieve analyse. Wageningen, A.A.G. Bijdragen 31. (*The economic development of agriculture in Groningen 1800-1910 ...*).

Priester, P.R. – 1998. Geschiedenis van de Zeeuwse landbouw. Wageningen, A.A.G. Bijdragen 37. (*History of agriculture in Zeeland*).

Quanjer, H.M. – 1911. Het besproeien der aardappelen met Bordeauxsche pap. Den Haag, Van Langenhuysen. (*Spraying potatoes with Bordeaux mixture*).

Quanjer, H.M. – 1912. Iets over de techniek van het sproeien. Tijdschrift over Plantenziekten 18: 55-60. (*Something on the technique of spraying*).

Quanjer, H.M. – 1913. Die Nekrose des Phloëms der Kartoffelpflanze, die Ursache der Blattrollkrankheit. Mededeelingen van de Rijks Hoogere Land-, tuin- en boschbouwschool 6: 42-80, with illustrations. (*The phloem necrosis of the potato plant, the cause of the leaf curl disease*).

Quanjer, H.M. – 1914. Iets over de techniek van het sproeien (Vervolg). Tijdschrift over Plantenziekten 20: 28-35. (*Something on the technique of spraying (Continuation)*).

Quanjer, H.M. – 1921a. Wetenschappelijk onderzoek en Regeringszorg voor de aardappelcultuur. Verslagen en Mededeelingen van de Directie van den Landbouw, 1921, No 1: 7-25. (*Scientific research and governmental concern for the potato cultivation*).

Quanjer, H.M. – 1921b. New work on leaf-curl and allied diseases in Holland, p. 127-145. In: W.R. Dykes (Ed.) Report of the International Potato Conference. London, Royal Horticultural Society.

Quanjer, H.M., Dorst, J.C., Dijt, M.D., Van der Haar, A.W. – 1920. De mozaïekziekte van de Solanaceeën, hare verwantschap met de phloeemnecrose en hare beteekenis voor de aardappelcultuur. Mededeelingen van de Landbouwhoogeschool 17: 1-74. (*The mosaic disease of the Solanaceae, her relationship to the phloem necrosis and her importance for the potato cultivation*).

Quanjer, H.M., Van der Lek, H.A.A., Oortwijn Botjes, J. – 1916. Aard, verspreiding en bestrijding van Phloeemnecrose (bladrol) en verwante ziekten, o.a. Sereh. Mededeelingen van de Rijks Hoogere Land-, Tuin- en Boschbouwschool 10: 1-162, with plates and figures. (*Nature, mode of dissemination and control of phloem-necrosis and related diseases*).

Quételet, L.Ad.J. – 1846. Phénomènes périodiques. Bulletin de l'Académie royale des sciences, des lettres et des beaux-arts de Belgique 13 (I): 162-165. (*Periodical phenomena*).

Rabl, M. (Ed.) – 1906. Rudolf Virchow. Briefe an seine Eltern. 1839 bis 1864. Leipzig, Engelman. (*Rudolf Virchow. Letters to his parents. 1839 to 1864*).

Rackham, H. – 1965. Aristotle. Problems II. London, Heinemann.

Rackham, H. – 1971. Pliny. Natural history. Volume V. Libri XVII-XIX. London, Heinemann.

Rapilly, F. – 1979. Yellow rust epidemiology. Annual Review of Phytopathology 17: 59-73.

Rapport – 1845. Rapport adressé à M. le ministre de l'intérieur par la Commission chargée de l'examen de questions relatives la maladie des pommes de terre. Le Moniteur Belge 18: 2422-2424, 2513-2515, 2693-2694. (*Report to the Home Secretary by the committee charged with the examination of questions relating to the potato disease*).

Reader, J. – 2008. Propitious esculent. The potato in world history. London, Heinemann.

Real, W. – 1983. Die Revolution in Baden 1848/49. Stuttgart, Kohlhammer. (*The revolution in Baden 1848/49*).

Regel, E. – 1854. Die Schmarotzergewächse und die mit denselben in Verbindung stehenden Pflanzen-krankheiten. Zürich, Schulthetz. (*Parasitic growths and the related plant diseases*).

Remarque, E.M. – 1929. Im Westen nichts Neues. Berlin, Propyläen-Verlag. (*No news from the West*).

Remy, Th. – 1916. Sorte und Saatgut in ihrer Bedeutung für den Ausfall der Kartoffelernte. Mitteilungen der deutschen Landwirtschafts-Gesellschaft 32: 814-822. (*Variety and planting material in their impact on the failure of the potato harvest*).

Reynebeau, M. – 2005. Een geschiedenis van België. 7th Ed. Tielt, Lannoo. (*A history of Belgium*).

Riegger, W. – 1998. Die Badische Revolution 1848/49. (*The revolution in Baden 1849/49*) www.junggesellen. de/archiv_heimatabend98.htm (last consulted July, 2005).

Rikli, M. – 1946. Das Pflanzenkleid der Mittelmeerländer. 2nd Ed. Vol II. Bern, Huber.(*The plant cover of the Mediterranean countries*).

Ritter, G. – 1917. Die phänologischen und meteorologischen Verhältnisse des Jahres 1916. Fühlings Landwirtschaftliche Zeitung 66: 123-127. (*The phenological and meteorological conditions of the year 1916*).

Ritzema Bos, J. – 1919. Bijdrage tot de kennis van de werking der Bordeauxsche pap op de aardappelplant. Tijdschrift over Plantenziekten 25: 77-94. (*Contribution to the knowledge of the effect of Bordeaux mixture on the potato plant*).

Roelfs, A.P. – 1985. Wheat and rye stem rust, p. 4-37. In: A.P. Roelfs & W.R. Bushnell (Eds.), The cereal rusts II. Orlando, Academic Press.

Roessingh, H.K., Schaars, A.H.G. – 1996. De Gelderse landbouw omstreeks 1825. Maastricht, Vereniging voor Landbouwgeschiedenis. (*Agriculture in Gelderland around 1825*).

Rosenzweig, I. – 1937. Ritual and cults of pre-Roman Iguvium. London, Christophers.

Rossikow, K.W. – 1916. Über die Feldmausplage und die natürlichen Ursachen ihres plötzlichen Verschwindens im Distrikt Uman Provinz Kiew im Jahre 1915. Landwirthschaftliche Zeitung, Petersburg 1916: 860-958. (*On the vole outbreak and the natural causes of its sudden disappearance in the district Uman province Kiev in the year 1915*).

Rostrup, E. – 1905. Oversigt over Landbrugsplanternes Sygdomme i 1904. Tidsskrift for Landbrugets Planteavl 2: 352-376. (*Overview of diseases on arable crops in 1904*).

Rostrup, E. – 1906. Oversigt over Landbrugsplanternes Sygdomme i 1905. Tidsskrift for Landbrugets Planteavl 3: 79-10. (*Overview of diseases on arable crops in 1905*).

Roze, E. – 1898. Histoire de la pomme de terre, traitée aux points de vue historique, biologique, pathologique, cultural et utilitaire. Paris, Rothschild. (*History of the potato, treated from historical, biological, cultural and utilitarian points of view*).

Rozendaal, A. – 1949. De betekenis van Quanjer voor het virusonderzoek. Tijdschrift over Plantenziekten (European Journal of Plant Pathology) 55: 103-108. (*The significance of Quanjer for virus research*).

Rozier (Ed.) – 1809. Cours complet d'agriculture pratique, d'économie rurale et domestique, et de médicine vétérinaire. Paris, Buisson et al. (*Complete course in practical agriculture, rural and home economics, and veterinary medicine*).

Rumpler, H. – 1997. Österreichische Geschichte 1804-1914. Eine Chance für Mitteleuropa. Bürgerliche Emanzipation und Staatsverfall in der Habsburgmonarchie. Wien, Ueberreuter. (*Austrian history 1804-1914. An opportunity for Middle Europe ...*).

Rupprecht, XX. – 1847. Ansichten über die Kartoffelkrankheit. Oekonomische Neuigkeiten und Verhandlungen 71: 429-430. (*Views on the potato disease*).

Rüter, A.J.C. (Ed.) – 1950. Rapporten van de gouverneurs in de provinciën 1840-1849. Vol. 3. Periodieke rapporten 1844, 1845. Utrecht, Kemink. (*Reports of the governors in the provinces 1840-1849*).

Rüter, H. – 1980. Remarques Im Westen Nichts Neues. Ein Bestseller der Kriegsliteratur im Kontext. Paderborn, Schöning. (*Remarque's 'No news from the West'. A bestseller of the war literature in its context*).

Sajet, B.H., Polak, W. – 1916. Eene voedings-enquête in den mobilisatietijd. Amsterdam, Ontwikkeling. (*An enquiry into nutrition in mobilization time*)

Salaman, R.N. – 1949. Some notes on the history of curl. Tijdschrift over Plantenziekten (European Journal of Plant Pathology) 55: 118-128.

Salaman, R.N. – 1970. The history and social influence of the potato. Cambridge, Cambridge University Press. Reprint from 1949.

Sandgruber, R. – 1978. Wirtschafts- und Sozialstatistik Österreich-Ungarns. I. Österreichische Agrarstatistik 1750-1918. Vienna, Verlag für Geschichte und Politik. (*Economic and social statistics of Austria-Hungary. I. Austrian agricultural statistics 1750-1918*).

Sauberg, F. – 1845. Die Kartoffelkrankheit im Jahre 1845. Cleve, Char. (*The potato disease in the year 1845*).

References

Savile, D.B.O. – 1984. Taxonomy of the cereal rust fungi, p. 79-112. In: W.R. Bushnell, A.P. Roelfs (Eds.), The cereal rusts, Vol. I. Orlando, Academic Press.

Savulescu, T. – 1953. Monografia uredinalelor din Republica Populară Română. Vol. I & II. Bukarest, Editura Academiei. (*Monography of the rusts in the Romanian People's Republic*).

Schacht, H. – 1856. Bericht an das Königliche Landes-Oekonomie-Collegium über die Kartoffelpflanze und deren Krankheiten. Berlin, Wiegandt. (*Report to the Royal National-economic Council on the potato plant and her diseases*).

Schander, R. – 1915. Bericht über die Tätigkeit der Agrikultur-botanischen Versuchs- und Samenkontroll Station von der Landkammer für die Provinz Schlezien zu Breslau vom 1 April 1914 bis 31 März 1915. Abstract in Zeitschrift für Pflanzenkrankheiten und Pflanzenschutz 25 (1915) 126. (*Communication on the activities of the agricultural-botanical experiment and seed-testing station ... Silesia ...*).

Schander, R. - 1917. Die Kartoffelfehlernte 1916 und ihre Ursachen. Fühlings Landwirtschaftliche Zeitung 66 (7/8): 145-168. (*The failure of the potato harvest 1916 and its causes*).

Schander, R. – 1918. Beobachtungen und Versuche über Kartoffeln und Kartoffelkrankheiten im Sommer 1917. Fühlings Landwirtschaftliche Zeitung 67: 204-226. (*Observations and experiments on potato and potato diseases in the summer of 1917*).

Schander, R., Krause, F. – 1915. Zur Mäusefrage. Fühling's Landwirtschaftliche Zeitung 64: 215-232. (*On the problem of mice*).

Schander, R., Krause, F. – 1917. Berichte über Pflanzenschutz der Abteilung für Pflanzenkrankheiten des Kaiser Wilhems-Instituts für Landwirtschaft in Bromberg. Die Vegetationsperiode 1913/14. Berlin, 1916. Abstract in Zeitschrift für Pflanzenkrankheiten 27: 128-130. (*Reports on crop protection by the department of plant diseases of the Emperor Wilhelm Institute ... vegetation period 1913/14*).

Schander, R., Tiesenhausen – 1914. Mitteilungen der Abteilung für Pflanzenkrankheiten am Kaiser Wilhelms-Institut in Bromberg. Mitteilungen Bd VI Heft 2. Abstract in Zeitschrift für Pflanzenkrankheiten 25 (1915) 16-17. (*Reports of the Department for Plant Diseases...*).

Scheibe, A. – 1933. Die Schwarzrostepidemie auf dem Balkan 1932. Nachrichtenblatt für den Deutschen Pflanzenschutsdienst 13: 5-6. (*The stem rust epidemic on the Balkan 1932*).

Schick, R., Klinkowski, M. (Eds.) – 1962. Die Kartoffel, ein Handbuch. Vol II. Berlin, VEB Deutscher Landwirtschaftsverlag. (*The potato, a handbook*).

Schirm, J.W. – 1846. Ursachen, Wesen und Entwicklung der Kartoffelkrankheit im Jahre 1845, nebst einigen Andeutungen über Aufbewahrung und Fortpflanzung der Kartoffeln. Oekonomische Neuigkeiten und Verhandlungen 71: 297-303. (*Causes, nature and development of the potato disease in the year 1845, with some indications on storage and propagation of the potatoes*).

Schmidt, O. – 1917. Zur Kenntnis der durch Fusarien hervorgerufenen Krankheitserscheinungen der Halmfrüchte. Fühlings landwirtschaftliche Zeitung 66: 65-84. (*To the knowledge of the disease symptoms of cereals caused by Fusaria*).

Schneider, G. – 1918a. Behandlung und Pflege der Kartoffeln in den Aufbewahrungsräumen. Deutsche Landwirtschaftliche Presse 45: 481-483. (*Treatment and care of potatoes in the storage areas*).

Schneider, G. – 1918b. Kartoffel-Lagerschuppen und –Lagerhäuser. Deutsche Landwirtschaftliche Presse 45: 536-537. (*Potato storage barns and houses*)

Schneider, G. – 1918c. Das Faulen der eingelegten Kartoffeln und seine Verhinderung durch sachgemäsze Durchlüftung und Durchkühlung der Aufbewahrungsräume. Deutsche Landwirtschaftliche Presse 45: 599. (*The rotting of stored potatoes and its prevention by appropriate aeration and cooling of the storage rooms*).

Schneiderhan, F.J. – 1933. The discovery of Bordeaux Mixture. Phytopathological Classics 3: 1-25.

Schulz, XX. – 1846. Berichte. Preuszisch-Schlesien. Oekonomische Neuigkeiten und Verhandlungen 71: 191. (*Reports. Prussian-Silesia*).

On the political economy of plant disease epidemics

Schwarz, M. – 1919. Über die Ausbreitung der Feldmäuse in Deutschland im Sommer und Herbst 1918. Fühling's Landwirtschaftliche Zeitung 68: 476-477. (*On the increase of field mice [voles] in Germany in the summer and fall of 1918*).

Schwarz, M. – 1920. Mäuseplage in Deutschland im Jahre 1919. Berichte über die Tätigkeit der Biologischen Reichsanstalt für Land- und Forstwirtschaft im Jahre 1919. Berlin, Parey: 71-80. (*Mice outbreaks in Germany in the year 1919*).

Seelman-Eggebert, E.L. – 1928. Friedrich Wilhelm Raiffeisen, sein Lebensgang und sein genossenschaftliches Werk. Stuttgart, Kohlhammer. (*F.W. Raiffeisen, his life and his work on co-operatives*).

Semal, J. – 1995. L'épopée du mildiou de la pomme de terre (1845-1995). Cahiers Agriculture 4: 287-298. (*The epic of potato late blight (1845-1995)*).

Sen, A. – 1981. Poverty and famines. An essay on entitlement and deprivation. Oxford, Clarendon.

Sen-Gupta, D.N. – 1945. Famine prevention works in Bengal. The Asiatic Review 41: 65-72.

Seringe, N.C. – 1818. Monographie des céréales de la Suisse. Bern, s.n. (*Monography of the cereals of Switzerland*).

Shapiro, S.S., Francia, R.S. – 1972. An approximate analysis of variance test for normality. Journal American Statistical Association 67 (337) 215-216.

Siney, M.C. – 1957. The allied blockade of Germany, 1914-1916. Ann Arbor, University of Michigan.

Slicher Van Bath, B.H. – 1987. De agrarische geschiedenis van West-Europa 500-1850. 6th Ed. Utrecht, Spectrum. (*The agrarian history of Western Europe 500-1850*).

Smit, C. – 1973. Nederland in de Eerste Wereldoorlog (1899-1919) III. 1917-1919. Groningen, Wolters-Noordhoff. (*The Netherlands in the First World War (1899-1919) III. 1917-1919*)

Sneller, Z.W. (Ed.) – 1943. Geschiedenis van den Nederlandschen landbouw 1795-1940. Groningen, Wolters. (*History of Dutch agriculture 1795-1940*).

Soetens, C. – 1834. Landbouw. Aardappelenzaad. Wetenschappelijk Maandschrift 1: 223-226. (*Agriculture. Potato seed*).

Souheur, J. – 1892. Praktische handleiding der behandeling van den valschen meeldauw, brand, verrotting en de anthracnose van den wijngaard, tomaat, aardappel en de meeste vruchtboomen met 27 afbeeldingen door de koper-sulfosteatiet beste geneesmiddel tegen alle kryptogamische ziekten, enz. enz. Antwerpen. (*Practical manual for the treatment of downy mildew … of the vineyard, tomato, potato and most of the fruit trees with 27 pictures by copper-sulfosteatit best medicine against all cryptogamic diseases …*).

Spahr Van der Hoek, J.J., Postma, O. – 1952. Geschiedenis van de Friese landbouw. I & II. Uitg. Friesche Maatschappij van Landbouw. (*History of Frisian agriculture*).

Springer, A. – 1865. Geschichte Österreichs seit dem Wiener Frieden 1809. Leipzig, Hirzel. Vol. 2, 1865. (*Austria's history since the Peace of Vienna 1809*).

Staring, W.C.H. – 1860. Huisboek voor den landman in Nederland. Haarlem, Kruseman. (*House book for the husbandman in the Netherlands*).

Stevens, C.E. – 1942. Agriculture and rural life in the later Roman Empire, p. 89-117. In: J.H. Clapham & E. Power (Eds.), The Cambridge economic history of Europe. Cambridge, Cambridge University Press.

Struik, P.C., Wiersema, S.G. – 1999. Seed potato technology. Wageningen, Wageningen Pers.

Tauger, M.B. – 2003. Entitlement, shortage and the 1943 Bengal Famine: another look. The Journal of Peasant Studies 31: 45-72.

Terlouw, F. – 1971. De aardappelziekte in Nederland in 1845 en volgende jaren. Economisch en sociaal-historisch Jaarboek 34: 263-308. (*The potato blight in the Netherlands in 1845 and following years*).

Tessier, H.-A. – 1783. Traité des maladies des grains. Paris, Herissant & Barrois. (*Treatise on the diseases of cereals*).

References

Tessier, H.-A., Thouin, A., Bosc d'Antic, L.-A. - 1818. Encyclopédie méthodique, ou par ordre de matières, par une Société de gens de lettres, de savans et d'artistes. Paris, Panckoucke, 1777-1792 et Paris, Agasse, 1793-1832. Vol. 4, Part 5, p195. (*Methodical encyclopaedia...*).

Tiesing, H. – 1923. Zaaikoren. Cultura 35: 288-289. (*Corn seed for sowing*).

Tozzetti, G. Targioni – 1767. True nature, causes and sad effects of the rust, the bunt, the smut, and other maladies of wheat, and of oats in the field. Translated in: Phytopathological Classics 9 (1952).

Treviranus, L.C. – 1835/8. Physiology der Gewächse. Part 1, 1835; part 2 1838. Bonn, Marcus. (*Plant physiology*)

Treviranus, L.C. – 1846. Der Spelzenbrand im Roggen. Botanische Zeitung 4: 629-631. (*The glume blast in the rye*).

Tulasne, L.-R. – 1853. Mémoire sur l'ergot des Glumacées. Annales des Sciences Naturelles, Botanique, Troisième série 20: 5-56. (*Memoir on the ergot of grasses*).

Tulasne, L.-R. – 1854. Second mémoire sur les uredinées et les ustilaginées. Annales des Sciences Naturelles, Botanique, Quatrième série 7: 12-127. (*Second memoir on the rusts and smuts*).

Tull, I. (Jethro) – 1733. The Horse-Hoing Husbandry: or, an essay on the principles of tillage and vegetation. Wherein is shewn a method of introducing a sort of vineyard-culture into the corn-felds, in order to increase their product, and diminish the common expence; by the use of instruments described in cuts. London, Strahan *et al.*

Turner, R.S. – 2006. After the famine: Plant pathology, *Phytophthora infestans*, and the late blight of potatoes, 1845-1960. Historical Studies in the Physical and Biological Sciences 35: 341-370.

Uilkens, T.F. – 1852. Groot warmoezeniersboek inhoudende eene volledige beschrijving der planten die in de moestuin voorkomen. Leiden, Noothoven van Goor. (*Large vegetable growers' book containing a complete description of all plants growing in the vegetable garden*).

Uilkens, Th.F. – 1853. Tuin-almanak, of de nieuwe opregte hollandsche hovenier; aanwijzende wat men 's maandelijks in moes- en bloemtuin, boomgaard, boomkwekerij, oranjerie en broeijerij te verrigten hebben. Gouda, Van Goor. (*Garden almanach ... indicating the monthly actions to be taken in vegetable and flower garden ...*).

Uilkens, Th. F. – 1855. Geschiedenis van den tuinbouw in Nederland, en overzigt van die in de verschillende staten van Europa. Groningen, J.B. Wolters. *(History of horticulture in the Netherlands, and overview of those in other states of Europe).*

Unger, F. – 1833. Die Exantheme der Pflanzen und einige mit diesen verwandten Krankheiten der Gewächse pathogenetisch und nosografisch dargestellt. Wien, Gerold. (*The exanthemes [eruptions, rashes] of plants and some related crop diseases presented pathogenetically and nosographically*).

Unger, D.F. – 1847. Botanische Beobachtungen. IV. Beitrag zur Kenntnis der in der Kartoffelkrankheit vorkommende Pilze und der Ursache ihres Entstehens. Botanische Zeitung 5: 305-317 + Figure 6. (*Botanical observations. IV. Contribution to the knowledge of the fungi occurring in the potato disease and the causes of their origin*).

Uppal, J.N. – 1984. Bengal Famine of 1943: A man-made tragedy. Delhi, Atma Ram.

Urban, Z. – 1967. The taxonomy of some European graminicolous rusts. Česká Mykologie 21: 12-16.

Urban, Z., Marková, J. – 1983. Ecology and taxonomy of *Puccinia graminis* Pers. in Czechoslovakia. Česká Mykologie 37: 129-150.

Valvo, A. – 1998. La 'Profezia di Vegoia': Proprietà fondiaria e aruspicina in Etruria nel I secolo A.C. (Studi pubblicati dall'Istituto Italiano per la Storia Antica, Vol. 43). Roma, p3, line 16. ('*The prophecy of Vegoia*' ...).

Van Aelbroeck, J.-L. – 1830. L'agriculture pratique de la Flandre. Paris, Huzard. (*The practical agriculture of Flanders*).

Van Bavegem, P.J. – 1782. Prijsverhandeling over de ontaarding der aardappelen. Dordrecht, Blussé. (*Price contest treatise on the degeneration of potatoes*).

Van der Hardt Aberson, F.E.C. – 1893. Het toenemen der ziekten in de cultuurgewassen, een dreigend gevaar voor de maatschappij. Leiden, Van Doesburgh. (*The increase of the diseases of cultivated crops, a menacing danger for the society*).

Van der Heiden, C. – 2001. De aardappelmisoogsten (1845-1848). Nieuwe Rotterdamsche Courant, 26 September. (*The potato crop failures (1845-1848)*).

Van der Meer, LB. – 2004. Etruscan origins. Language and archeology. Bulletin van de antieke beschaving (Leiden) 79: 51-57.

Vanderplank, J.E. – 1949. Some suggestions on the history of potato virus X. Journal of the Linnean Society, Botany 53: 251-262.

Vanderplank, J.E. – 1963. Plant diseases: epidemics and control. New York, Academic Press.

Vanderplank, J.E. – 1968. Disease resistance in plants. New York, Academic Press.

Van der Poel, J.G.M. – 1967. Honderd jaar landbouwmechanisatie in Nederland. Haarlem, Boeke & Huidekoper. (*Hundred years of agricultural mechanization in the Netherlands*).

Van der Wal, A.F., Zadoks, J.C. – 1971. Interaction of fungal pathogens, and its effect on crop losses in wheat. Proc. 2nd International Symposium Plant Pathology, Delhi.

Van der Zaag, D.E. – 1999. Die gewone aardappel. Geschiedenis van de aardappel en de aardappelteelt in Nederland. Wageningen, s.n. (*That ordinary potato; history of the potato and potato cultivation in the Netherlands*).

Van Ewyck, D.J. – 1847. Verslag van den Gouverneur en de Gedeputeerde Staten der Provincie Noord-Holland, aan de Provinciale Staten, in derzelve gewone vergadering, gehouden te Haarlem den 7[den] Julij 1846. (*Report of the Governor and the Board of the Province Noord-Holland to the Provincial States*).

Van Genderen, H., Schoonhoven, L.M., Fuchs, A. – 1996. Chemisch-ecologische flora van Nederland en België. Utrecht, KNNV. (*Chemical-ecological flora of the Netherlands and Belgium*).

Van Hall, H.C. – 1828. Gedachten over den honingdauw. Bijdragen tot de Natuurkundige Wetenschappen 3: stuk 1: 303-319. (*Thoughts on honeydew*).

Van Hall, H.C. – 1854. Neêrlands plantenschat, of landhuishoudkundige flora, behelzende eene beschrijving der onkruiden, vergiftige en nuttige inlandsche planten en der in onzen landbouw gekweekte gewassen. Leeuwarden, Suringar. (*Dutch plant treasure, or agricultural flora containing a description of weeds, poisonous and useful native plants and of crops grown in arable agriculture*).

Van Hall, H.C. – 1859. Honingdauw. Album der Natuur: 129-136. (*Honeydew*). Naschrift (*post-script*) P. Harting (p. 136-140).

Vanhaute, E. – 1992. Heiboeren. Bevolking, arbeid en inkomen in de 19[de] eeuwse Kempen. Brussel, VUBpress. (*Heathland farmers; population, labour and income in 19th century Campine*).

Van Oirschot, A., Jansen, A.C., Koesen, L.S.A. (Eds) – 1985. Encyclopaedie van Noord-Brabant. Vol. 1. Baarn, Market Books. (*Encyclopedia of North-Brabant*).

Van Peyma, W. – 1845. 'Letter to the editor'. Berigten en mededeelingen door het Genootschap voor Landbouw en Kruidkunde te Utrecht 3: 12-21.

Van Rossum, J.P. – 1845. 'Letter to the editor'. Berigten en mededeelingen door het Genootschap voor Landbouw en Kruidkunde te Utrecht 3: 5-9.

Van Schie, P. – 2006. The strengths and weaknesses of Dutch liberalism: a historical comparison with German liberalism, p. 35-53. In: P. Van Schie & G. Voerman (Eds.), The dividing line between success and failure. A comparison of liberalism in the Netherlands and Germany in the 19th and 20th centuries. Berlin, LIT Verlag.

Van Zanden, J.L. – 1985. De economische ontwikkeling van de Nederlandse landbouw in de negentiende eeuw, 1800-1914. Wageningen, A.A.G. Bijdragen 25. (*The economic development of Dutch agriculture in the nineteenth century, 1800-1914*).

References

Van Zanden, J.L. – 1991. 'Den zedelijken en materiëlen toestand der arbeidende bevolking te platten lande'. Een reeks rapporten uit 1851. Historia agriculturae 21. (*'The moral and material situation of the working population on the countryside', a set of reports from 1851*).

Varro – see Hooper, W.D. & Boyd Ash, H. (1967).

Verbruggen, J. – 1957. Berne handmissaal voor zondagen en feesten. Turnhout, Proost. (*Berne hand missal for Sundays and feasts*).

Vigreux, M. – 1998. Paysans et notables du Morvan au XIXe siècle jusqu'en 1914. Château-Chinon, Académie du Morvan. (*Peasants and dignitaries of the Morvan in the 19th century up to 1914*).

Virchow, R. – 1848. Mittheilungen über die in Oberschlesien herrschende Typhus-Epidemie. Berlin, Reimer. (*Communications on the typhus epidemic raging in Upper Silesia*).

Vis, C. – 1845. 'Letter to the editor'. Berigten en mededeelingen door het Genootschap voor Landbouw en Kruidkunde te Utrecht 3: 2-3.

Vissering, S. – 1845. Eenige opmerkingen ter zake van de aardappelziekte. Amsterdam, van Kampen (*Some comments concerning the potato disease*).

Vocke, R. – 1984. Der Erste Weltkrieg, pp 341-373 in H. Pleticha (Ed.): Deutsche Geschichte, Vol. 10: Bismarck-Reich und Wilhelminische Zeit. 1871-1918. Gütersloh, Bertelsmann. (*The First World War*).

Von Bujanovics, E. – 1847. Brot aus Quecken (*Triticum repens*). Oekonomische Neuigkeiten und Verhandlungen 73: 129-131. (*Bread from couch grass* (Triticum repens)).

Von Martius, C.F.P. - 1842. Die Kartoffel-Epidemie der letzten Jahre, oder die Stockfäule und Räude der Kartoffeln. München, s.n. (*The potato epidemic of the last years, or the storage rot and the scab of the potatoes*).

Von Schlechtendal, D.F.L. – 1846. Kurze Notizen. Botanische Zeitung 4: 504. (*Short notes*).

Von Thümen, F. – 1886. Die Bekämpfung der Pilzkrankheiten unserer Culturgewächse. Versuch einer Pflanzentherapie. Wien, Faesy. (*The control of the fungal diseases of our crops. Attempt at a plant therapy*).

Von Viebahn, XX. – 1848. Bericht des Geheimen Ober-Finanz-Raths v. Viebahn über eine agronomisch-technologische Reise in den Provinzen Posen, Preuszen und Pommern vom 9. Juni bis 11. Juli 1846. Annalen der Landwirthschaft 6 (Vol. 11) 1-74. (*Report of ... v. Viebahn on an agronomical-technological visit to the Provinces Posen, Prussia and Pommerania ...*).

Von Wahl, C., Müller, K. – 1915. Bericht der Hauptstelle für Pflanzenschutz in Baden für das Jahr 1914. Stuttgart, Ulmer. Abstract in Zeitschrift für Pflanzenschutz 26 (1916) 196-197. (*Report of the main station for plant protection in Baden ... 1914*).

Vrolik, G. – 1845. Waarnemingen en proeven over de onlangs geheerscht hebbende ziekte der aardappelen, dd 15-11-1845. Voordracht Vergadering 1e Klasse Koninklijke-Nederlandsch Instituut van Wetenschappen, Letterkunde en Schoone Kunsten. Amsterdam, Sulpke. (*Observations and experiments on the recently prevailing disease of potatoes*)

Vrolik, G., Numan, A., Van Hall, H.C., Brandts, A. – 1846. Omtrent de uitkomsten, welke de zaaijing van het door het department van Binnenlandsche Zaken in den aanvang van 1846, aan de Klasse gezonden aardappelzaad heeft opgeleeverd. Verslagen, bij de Eerste Klasse Koninklijke-Nederlandsch Instituut van Wetenschappen, Leterkunde en Schoone Kunsten. (Ook in Nederlandsche Staatscourant). (*About the results that the sowing of the potato seeds, sent to the Class by the Department of the Interior in the beginning of 1846, has produced*).

Wahl, I., Anikster, Y., Manisterski, J., Segal, A. – 1984. Evolution at the center of origin, p. 39-77. In: W.R. Bushnell & A.P. Roelfs (Eds.), The cereal rusts Vol. I. Orlando, Academic Press.

Wahlberg, P.F. – 1847. Bidrag till kännedommen om potäternas sjukdom i Sverige åren 1845 och 1846. Stockholm, Elmen & Granberg. (*Contribution to the knowledge of the potato disease in Sweden in the years 1845 and 1846*).

On the political economy of plant disease epidemics

Weber, XX. – 1847. Über die dieszjährige Fruchternte und deren Ausfall in den verschiedenen, besonders teutschen Ländern, nach sichern und zuverlässigen Nachrichten. Oekonomische Neuigkeiten und Verhandlungen 73: 2-7, 12-15, 22-24. (*On this year's crop harvests and their losses in the various, especially German, countries, according to certain and reliable informations*).

Wehler, H.-U. – 1987. Deutsche Gesellschaftsgeschichte. II. Von der Reformära bis zur industriellen und politischen ,Deutschen Doppelrevolution' 1815-1845/49. 2nd Ed. München, Beck. (*German social history. II. ...*).

Wehnelt, B. – 1943. Die Pflanzenpathologie der deutschen Romantik als Lehre vom kranken Leben und Bilden der Pflanzen. Bonn, Bonner Universitäts-Buchdruckerei. (*The phytopathology of German Romanticism as the theory of diseased life and formation of plants*).

Wennink, C.S. – 1918. De gevolgen der bladrolziekte bij aardappelen. Tijdschrift over Plantenziekten 24: Bijblad 1-4. (*The consequences of leaf roll in potatoes*).

Westendorp, G.D. – 1854a. Quatrième notice sur quelques cryptogames récemment découvertes en Belgique. Bull. Académie Royale de Sciences de Belgique 21: 229-246. (*Fourth note on some cryptogames recently discovered in Belgium*).

Westendorp, G.D. – 1854b. Description de quelques cryptogames inédites ou nouvelles pour la flore des deux Flandres. Académie Royale de Bruxelles (Extrait du tome XII, no 9, des Bulletins). (*Description of some cryptogames unpublished or new to the flora of the two Flanders*).

Westendorp, G.D. – 1854c. II. Notice sur quelques cryptogames inédites ou nouvelles pour la flore belge. Académie Royale de Belgique. (Extrait du tome XVIII, nos 7 et 10 des Bulletins de l'Académie Royale de Belgique). (*Note on some cryptogames unpublished or new to the Belgian flora*).

Westendorp, G.D. – 1860. Septième notice sur quelques cryptogames inédites ou nouvelles pour la flore belge. Extrait des Bulletins de l'Académie Royale de Belgique, 2me série, tome XI, no. 6. (*Seventh note on some cryptogames unpublished or new to the Belgian flora*).

Westendorp, G.D., Wallays, A.C.F. – 1845/50. Herbier cryptogamique ou collection de plantes cryptogames et agames qui croissent en Belgique. 28 fascicules in 18 'books' (*Cryptogamic herbal or collection of cryptogamic and agamic plants growing in Belgium*). Fascicules 1-5, 1845-1846. Bruges, de Pachtere. Fascicules 6-10, 1847-1850. Courtrai, Jaspin. Tableau méthodique, index of first 10 fascicules, Courtrai, Jaspin, 1850.

Westendorp, G.D., Wallays, A.C.F. – 1855. Herbier cryptogamique [belge] ou collection des plantes cryptogames qui croissent en Belgique. Fascicules 21 & 22. Gand, Van Doosselaere.

Westerdijk, J. – 1910. Die Mosaikkrankheit der Tomaten. Mededelingen Phytopathologisch Laboratorium 'Willie Commelin Scholten' 1: 1-22. (*The mosaic disease of tomatoes*).

Westerdijk, J. – 1916. Die Mosaikkrankheit der Kartoffelpflanze. Jahresbericht der Vereinigung für Angewandte Botanik 14: 145-149. (*The mosaic disease of the potato plant*).

Westerdijk, J. – 1917. De nieuwe wegen van het phytopathologisch onderzoek. Amsterdam, de Bussy. (*The new avenues of phytopathological research*).

Westermann, G. – 1956. Grosser Atlas zur Weltgeschichte. Braunschweig, Westermann. (*Large atlas on world history*).

White, K.D. – 1970. Roman farming. Ithaca (NY), Cornell University Press.

Wiegmann, A.F. – 1839. Die Kranheiten und krankhaften Miszbildungen der Gewächse, mit Angabe der Ursachen und der Heilung oder Verhütung derselben, so wie über einige den Gewächsen Schädliche Thiere und deren Vertilgung. Braunschweig, Viehweg. (*The diseases and morbid deformations of the crops, ...*).

Wiese, M.V. – 1977. Compendium of wheat diseases. St Paul, American Phytopathological Society.

Windt, L.G. – 1806. Der Berberitzenstrauch, ein Feind des Wintergetreides. Aus Erfahrungen, Versuchen und Zeugniszen. Bückeburg, s.n. / Hannover, Hahn. (*The barberry shrub, a foe of the winter corn [= rye]; from experiences, experiments and testimonies*).

Woodham-Smith, C. – 1962. The Great Hunger. Ireland 1845-1849. London, Hamilton.

Woodward, E.L. – 1938. The age of reform 1815-1870. Oxford, Clarendon.

Wttewaaall, J. – 1848. De roest der rogge. Algemene landhuishoudelijke Courant, Arnhem, Meijer. 2-25: Zaterdag 17 Juni. (*The rust of the rye*).

Wumkes, G.A. – 1934. Stads- en dorpskroniek van Friesland II (1800-1900). Leeuwarden, Eisma. (*Town and village chronicle of Friesland*).

Wyckoff, P. – 1972. Wallstreet and the stock markets. A chronology. 1644-1971. Philadelphia, Chilton.

IJnsen, F. – 1976. De zomers in Nederland vanaf 1706 thermisch bekeken. KNMI Wetenschappelijk Rapport W.R. 76-15. De Bilt, KNMI. (*The summers in the Netherlands from 1706, from a thermal viewpoint*).

IJnsen, F. – 1981. Onderzoek naar het optreden van winterweer in Nederland. KNMI Wetenschappelijk Rapport W.R. 74-2. 2nd Ed. De Bilt, KNMI. (*Study on the occurrence of winter weather in the Netherlands*).

Zadoks, J.C. – 1961. Yellow rust on wheat, studies in epidemiology and physiologic specialization. Wageningen, Veenman. Also in: Tijdschrift over Plantenziekten (European Journal of Plant Pathology) 67: 69-256.

Zadoks, J.C. – 1965. Epidemiology of wheat rusts in Europe. FAO Plant Protection Bull. 13: 97-108.

Zadoks, J.C. – 1984a. Cereal rusts, dogs, and stars in antiquity. Garcia de Orta, Ser. Estacão Agronómica Lisboa 9: 13-29.

Zadoks, J.C. – 1984b. Disease and pest shifts in modern wheat cultivation, p. 237-244. In: E.J. Gallagher (Ed.), Cereal production. London, Butterworths.

Zadoks, J.C. – 2003. Two wheat septorias; two emerging diseases from the past, p. 1-12. In: G.H.J. Kema, M. Van Ginkel & M. Harrabi (Eds.), Global insights into the *Septoria* and *Stagonospora* diseases of cereals. Proceedings of the 6th International Symposium on *Septoria* and *Stagonospora* Diseases of Cereals, December 8-12, 2003, Tunis, Tunesia.

Zadoks, J.C. – 2007. Fox and fire, a rusty riddle. Latomus (Brussels) 66: 3-9.

Zadoks, J.C. – 2008. The potato murrain on the European Continent and the revolutions of 1848. Potato Research 51: 5-45.

Zadoks, J.C., Chang, T.T., Konzak, C.F. – 1974. A decimal code for the growth stages of cereals. Weed Research 14: 415-421.

Zadoks, J.C., Leemans, A.M. – 1984. Bruine roest op tarwe in 1983; een verregende epidemie? Gewasbescherming 15: 5 (Abstract). (*Brown rust on wheat in 1983; an epidemic washed away by rain?*)

Zadoks, J.C., Schein, R.D. – 1979. Epidemiology and plant disease management. New York, Oxford University Press.

Zadoks, J.C., Van den Bosch, F. – 1994. On the spread of plant disease: A theory on foci. Annual Review of Phytopathology 32: 503-521.

Zimmermann, H. – 1916a. Bericht der Hauptsammelstelle Rostock für Pflanzenschutz in Mecklenburg im Jahre 1914. Abstract in Zeitschrift für Pflanzenkrankheiten 26: 387-388. (*Report of the main station Rostock for plant protection in Mecklenburg in the year 1914*).

Zimmermann, H. – 1916b. Eine Wurzelerkrankung des Roggens infolge Frostes. Zeitschrift für Pflanzenkrankheiten 26: 321-323. (*A root disease of rye due to frost*).

Curriculum vitae

Jan C. Zadoks was born in Amsterdam, 1929. He studied biology at the University of Amsterdam. He graduated in 1957, when he was a research officer at the Institute for Plant Protection Research (IPO-DLO), Wageningen. He received his Ph.D. from the University of Amsterdam in 1961, with honors, on a thesis 'Yellow rust on wheat, studies in epidemiology and physiologic specialisation'. He is married, and has four grown-up children and five grandchildren.

In 1961, Jan Zadoks joined the Wageningen Agricultural University. He became full professor of ecological plant pathology in 1969. He served 6 years as the honorary secretary of the Netherlands Phytopathological Society, 2 years as the secretary of the University Curriculum Committee, 3 years as the Dean of the Wageningen Agricultural University, and 2 years as Vice-President + 2 years as President of the Biology Section of the Netherlands Science Foundation (NWO). He served 3 years in the Pesticides Registration Board of The Netherlands. For 10 years he was a member of the Committee on Genetic Modification COGEM ('NGO Release Committee') of The Netherlands, with 5 years as chairman of the Subcommittee on Genetically Modified Plants.

He developed what was possibly the world's first course with practical in 'Plant disease epidemiology' and also courses in 'Aerobiology', 'Crop loss', 'Genetics of resistance' and 'Plant protection and society'. The first course led to Zadoks & Schein's 1979 book 'Epidemiology and plant disease management'. He initiated several (inter)national post-graduate courses on dynamic simulation in crop protection. Several post-graduates spent a sabbatical period with him. He lectured in many countries and presented invitational key-note lectures in various assemblies.

He did research in stripe rust, leaf rust, glume blotch and speckled leaf blotch of wheat. His 1974 scale for growth stages of cereals ('Decimal Code') became UPOV and FAO standard. He developed dynamic simulation in plant disease epidemiology, and initiated the development of the computerised pest and disease warning system EPIPRE for wheat. Later, he was involved in field studies, computer simulations and mathematical analyses of focus formation in plant disease. He took an interest in the development of 'alternative' agriculture and edited the 1989 booklet 'Development of farming systems, evaluation of the five-year period 1980-1984'. He (co-)authored over 400 papers. He supervised over 40 Ph.D. theses and he served repeatedly as an overseas external examiner.

Jan Zadoks had a strong interest in international agriculture. He founded the 'European and Mediterranean Cereal Rusts Foundation' in 1969. He performed consultancy missions overseas for FAO and for the Dutch and French governments (crop loss, resistance, IPM, teaching). He was a Scientific Councillor to the French overseas research organisations ORSTOM in France and IIRSDA in Ivory Coast. He participated in quinquennial reviews of DFPV, ICRISAT, IPO, IRHO, IRRI and ITC. For 14 years, he was a member of the FAO/UNEP Panel of Experts for Integrated Pest Control. He was a visiting professor at the University of Paris, France. He organised the XIIIth International Plant Protection Congress, 1995, The Netherlands.

He received the 'Adventurers in Agricultural Science Award of Distinction', Washington (1979), two Dutch Royal Awards for Public Merit (1980, officer in the 'Order of Orange Nassau'; 1993, knight in the 'Order of the Netherlands Lion'), and the Biannual Award of the Royal Netherlands Phytopathological Society (2002). He was appointed a 'Fellow' of the American Phytopathological Society in 1994, and he received an 'honorary doctorate' in agriculture from the Swedish Agricultural University, 2005.

Notes

Chapter 1

[1] E.g. the famine due to the siege of Jerusalem by Nebuchadnezzar, 2 Kings 25: 3 'And on the ninth day of the fourth month the famine prevailed in the city, and there was no bread for the people on the land.'

[2] E.g. locusts, one of the plagues of Egypt, Exodus 10: 12.

[3] E.g. Amos 4: 9 'I have smitten you with blasting and mildew: ...'

[4] The present paper is a compilation with amendments of two earlier papers (Zadoks 1984a, 2007).

[5] Magic involves a specific human act performed to obtain a specific material effect, in a one-to-one relation, thought to be a causal relation; e.g. a typical rain dance to obtain rain during a prolonged drought. Magic connects the earthly world with the transcendental world.

[6] Compare Diamond (1997). Many species of wheat exist, with one to three sets of chromosomes, i.e. from diploid to hexaploid. The term 'wheat' is used here for the cultivated crop, irrespective of species.

[7] For speciation the co-evolution of the rust with its alternate host, on which the rust's sexual stage occurs, was as important as the co-evolution with its grass host (Wahl et al., 1984; Anikster & Wahl, 1979). For the evolution of black rust and its adaptation to wheat see Urban (1967), Urban & Marková (1983) and Savile (1984).

[8] Macrocyclic – the stem rust fungus goes through an annual cycle with five spore forms. The rust overwinters in black pustules (telia) on wheat straw. Heteroecious – changing host plant species. In spring the telia produce teliospores that infect barberry bushes (Berberis vulgaris L.). On the barberry the rust produces first pycnidiospores and later aecidiospores (spring spores). The latter are wind-blown to the grass lands and wheat crops on which they cause fox-red pustules loaded with urediniospores (summer spores). This is the repetitive stage of the annual cycle with multiplication and dissemination, so that an epidemic may build up if conditions are favourable. The red pustules later turn black filling up with teliospores.
The wind-blown spring and summer spores can cover long distances (up to hundreds of kilometres) so that the sources of infection may remain unknown.
In the Mediterranean Area severe epidemics may appear without host plant alternation when the uredinial stage of the rust overwinters in pockets where suitable gramineous host plants, moisture and temperature concur.
A sentence by Pliny (NH XVIII. xlv. 161) can be read as suggesting a relation of the rust to barberries. 'As for the greatest curse of corn, mildew, fixing branches of laurel in the ground makes it pass out of the fields into their foliage'. (Rubigo quidem, maximum segetum pestis, lauri ramis in arvo defixis transit in ea folia ex arvis.) If the interpretation is correct, the rust was thought to pass from the fields into the barberry. Today we know that in the spring the rust passes from the barberry into the fields.

[9] Anikster & Wahl (1979: 372). Today's black rust may have evolved in an environment of open forests with grasslands by doubling its chromosome numbers and specialising as a tetraploid on wheat (Urban, 1967). The final word has not yet been said (Urban & Marková, 1983).

Notes

[10] 'Crop' should not be taken literally; self-sown wheat and grasses can also provide the bridge from one cropping season to the next.

[11] From W to E *Berberis hispanica* in the Atlas and the Spanish sierras, *B. aetnensis* on Corsica, Sardinia, Sicily and in Calabria, *B. cretica* on Crete and in Lebanon, and *B. crataegina* in Turkey and Azerbaijan. These taxa should be seen as subspecies of *B. vulgaris* (Rikli, 1946). They are low shrubs up to ~70 cm. Barberry rapidly colonises abandoned land. It may have been more common in antiquity than today since land use was less intensive.

[12] In Denmark barberry was planted extensively around 1800 for the 'division of the commons' with horrible consequences (Hermannsen, 1968). For the naughty boys see Oort (1941: 112).

[13] See e.g. Solomon's prayer in 1 Kings 8: 37 'If there be in the land famine, if there be pestilence, blasting, mildew, locust, *or* if there be caterpillar; ...' (King James Bible).

[14] Banks (1805) describing what is now called *P. graminis* f.sp. *tritici* distinguished rust (the uredinial stage) and mildew (the telial stage), as did the mycologists of that period.

[15] For the effect of black rust on rye stems the term 'chalk white' (G: *kalkweisz*) was also used (Windt, 1806: 9).

[16] Livid (Marshall, 1795: 359ff); '... a remarkably fine field of wheat ... as if it were covered with soot' (Lambert, 1789); Lambert's 'soot' quoted by Darwin (1800: 321). Field 'blasted' (Lambert, 1798).

[17] See field symptoms of black rust on rye in Windt (1806: 9, 25).

[18] Chester (1946); Hogg *et al.* (1969).

[19] Aristotle, Problems XXVI: 17, translation by Rackham (1965: 83).

[20] We cannot exclude the presence of brown leaf rust with rusty-brown pustules (uredinia) in addition to black stem rust with more fox-brown pustules. Descriptions of colour variants are notoriously unreliable (§2.2.2 and Appendix 2.1).

[21] Cato, Varro, Vergil, Columella, Ovid, Plinius, Palladius and others (Orlob, 1973: 127ff).

[22] Note the confusion. The final (= telial) stage of black stem rust is black, not brown or red.

[23] Pliny *Naturalis Historia* XVIII: 79. ... *hordeum ex omni frumento minime calamitosum, quia ante tollitur quam triticum occupet rubigo.* (Barley is the least liable to damage of all corn, because it is harvested before the wheat is attacked by mildew). Mildew is an old English term for black rust (telial stage). The barley harvest may precede the wheat harvest several weeks (§1.4.2).

[24] Pliny *Naturalis Historia* XVIII: 161.

[25] Pliny *Naturalis Historia* XVIII: 92. ... *sed minus quam cetera frumenta in stipula periclitatur, quoniam semper rectam habet spicam nec rorem continent qui robiginem faciat.* (But it [the wheat] is less exposed to danger in the straw than other cereals, because it always has the ear on a straight stalk and it does not hold dew to cause rust). In the modern eye this is a questionable statement.

Pliny *Naturalis Historia* XVIII: 154. ... *caeleste frugum vinearumque malum nullo minus noxium est robigo. Frequentissima haec in roscido tractu convallibusque ac perflatum non habentibus; e diverso carent ea ventosa et excelsa.* (One of the most harmful climatic maladies of corn crops and vines is rust. This is most frequent in a district exposed to dew and in shut-in valleys that have no current of air through them, whereas windy places and high ground on the contrary are free from it.) This is a correct statement. The 'rust' on vines refers to brown discoloration of vine leaves.

Pliny *Naturalis Historia* XVIII: 279.

See Hogg *et al.*, 1969. During clear nights following hot days, heavy dew may be formed, necessary for the germination of the black rust spores. The sequence of temperatures, cool nights and warm days, is just right for penetration and infection.

[26] The pustules were on the inner, concave side of a lemma. The uredeniospores with their four equatorial germpores clearly belonged to *P. graminis*. Tetraploid *T. parvicoccum* is sometimes included in *T. turgidum*.

[27] Example: During a prolonged drought people may perform a specific rain dance to invoke the much desired rain.

[28] Ovid, *Fasti* IV: 679 ff. The cereals were wheat and barley. Often, leguminous crops such as beans and peas were included in the cereals, seen as 'seed crops'.

[29] Ovid, *Fasti* IV: 681/2. *Cur igitur missae vinctis ardentia taedis terga ferant volpes,causa docenda mihi est.* (I must therefore explain the reason why foxes are let loose with torches tied to their burning backs.)

[30] Or through the land of Carseoli, a town near Rome; the text is not clear about the location.

[31] Ovid, *Fasti* IV: 683 ff.

[32] The Praenestine calender of Roman festivals was engraved in marble. Fragments were found in the Italian town of Praeneste, now Palestrina, SE of Rome (Wikipedia, English version).

[33] Missale romanum ex decreto sacrosancti concilii tridentni restitutum ..., Venice, Pezzana, 1756.

[34] See Bain *et al.* (1995: 33) on France.

[35] A feast, the Floralia, was held on April 28th in honour of the goddess Flora.

[36] Varro in 'On agriculture' I: i, 6 (Hooper & Boyd Ash, 1967): '... *Robigum ac Floram; quibus propitiis neque robigo frumenta atque arbores corrumpit ... Itaque publice Robigo feriae Robigalia ... instituti.*' (... Robigo and Flora; for when they are propitious the rust will not harm the grain and the trees ... wherefore, in honour of Robigus has been established the solemn feast of the Robigalia ...)

[37] Elsewhere also called Robigo. The gender of the deity varies, female and male possibly expressing malevolence and benevolence, respectively (Leopold, 1926).

[38] Varro, *De lingua Latina* VI §16 (Kent, 1967). *Robigalia dicta ab Robigo; secundum segetes huic deo sacrificatur, ne robigo occupet segetes.* (The *Robigalia* 'Festival of Robigus' was named from Robigus 'God of Rust'; to this god sacrifice is made along the cornfields, that rust may not seize upon the standing corn.)

[39] I have no doubt about the correctness of Leopold's (1926) statements on Roman antiquity but his biology was weak; he thought ergot and rust to be the same thing.

[40] The barley – wheat difference was mentioned by Pliny (*Naturalis Historia* XVIII: 79).

[41] Nazareno Strampelli (1866-1942), Italian plant breeder, created new wheat varieties (~1920/30) with early ripening, to escape from stem rust (Enciclopedia Italiana; Wikipedia). These varieties can be harvested one to two weeks before traditional varieties. The Broekhuizen (1969) data refer to the new varieties.

[42] Compare Tull (1733: 55) on wheat drilling 'if too thin, it may happen to tiller so late in the Spring, that some of the Ears may be blighted, yet a little thicker or thinner does not matter.'

[43] The three wheat rust species, yellow stripe rust (*Puccinia striiformis*), brown leaf rust (*P. triticina*) and black stem rust (*P. graminis*) tend to appear in this order. Stripe rust prefers lower and stem rust higher temperatures, whereas leaf rust has a broad intermediate temperature preference (see Box 2.1).

[44] An early reference is Marshall, 1795 (Minutes 13 and 133). See Hogg *et al.* (1969: 7).

[45] Such as Orlob (1973), Postan (1966), Slicher Van Bath (1987).

[46] Translation by R. Kotansky (1994).

[47] Clark (2007): 'The heliacal setting is the last day when the star [Sirius] sets and the sun is far enough below the western horizon to make the star visible in the evening twilight'. Follows a period of ~85 days during which Sirius is invisible. 'The heliacal rising is the first day when the star rises and the sun is far enough below the eastern horizon to make it visible in the morning twilight'. Precise dates vary with time and geographic position of the observer.

[48] Pliny *Naturalis Historia* XVIII: 288/9.

[49] Pliny *Naturalis Historia* XVIII: 290 'Within these periods falls the sterilising influence of the heavens, though I would not deny the possibility that it is liable to alteration by local climatic conditions ...'.

[49a] In Zadoks (2007) I wrongly called Sirius the 'morning star' in stead of 'evening star'.

[50] At Roman latitude the heliacal setting of Sirius took place when the sun was in the tenth degree of the Bull (Varro quoted by Pliny *Naturalis Historia* XVIII: 285). The ancients knew that the heliacal rising of Sirius took place when the sun rose in the sign of the Lion, a timing accepted by the Romans. Today the two stars rise simultaneously in the sign of Cancer, the heliacal rising of Sirius at the latitude of Rome being August 11th and its setting May 18th. The Cancer-Lion difference is due to the precession, the astronomical phenomenon of the slow but constant change in position of the equinoxes relative to the stars.

[51] Plinius *Naturalis Historia* XVIII: 285. *Robigalia Numa constituit anno regni sui XI, quae nunc aguntur a.d. VII kal. Mai., quoniam tunc fere segetes robigo occupat. ... canis occidit, sidus et per se vehemens ...*

[52] Pliny *Naturalis Historia* XVIII: 291: *et in hoc mirari benignitatem naturae succurrit: iam primum hanc iniuriam omnibus annis accidere non posse propter statos siderum cursus, ...*

[53] In the Bible, Solomon's Song 2: 15, damaging foxes are mentioned that cannot be related to wheat rust. 'Take us the foxes, the little foxes, that spoil the vines: ...'.

[54] I saw such restricted burnt areas in forests, Washington State, U.S.A., with areas measuring few ares. For dry wheat fires see http://emd.wa.gov/6-mrr/resp/seoo/stats-archive/stats-02/seoo-07-02.htm accessed January, 2006. Burnt areas covered many hectares.

[55] Professor Amos Dinoor from the Hebrew University kindly informed me so.

[56] In Southern Europe, intercropping of olives and wheat was frequent up to the 1950s at least (author's observation).

[57] Pliny *Naturalis Historia* XVIII: 275. The term 'rust' was used in its present specific meaning but also in a more general sense indicating brownish discolorations of leaves, e.g. by nightfrost.

[58] The Iguvine Tablets, made ~200 BCE, and containing a text from ~700 BCE were found in Gubbio, a town in Umbria, Italy (Rosenzweig, 1937; Devoto, 1948).

[59] Columella X: 342/3. *Hinc mala rubigo virides ne torreat herbas, sanguine lactentis catuli placatur et extis.* (Hence, lest fell Rubigo parch the fresh, green plants, her anger is appeased with blood and entrails of a suckling whelp: Forster & Heffner, 1968)

[60] Ovid (Frazer, 1989: 422): The dog sacrifice (*sacrum canaria*) near the Doggy Gate (*Porta Catularia*). Festus 45.10 (Lindsay, 1965): *Catularia porta Romae dicta est, quia non longe ab ea ad placandum caniculae sidus frugibus inimicum rufae canes immolabantur, ut fruges flavescentes at maturitatem perducerentur.* (It was said that there is a Doggy Gate at Rome, since not far away red-haired dogs were sacrificed to conciliate the Dogstar, hostile to the crops, so that the crops may yellow and attain full maturity).

[61] Pliny *Naturalis Historia* XVIII: 14. *Augurio canario agenda dies constituatur priusquam frumenta vaginis exeant et antequam in vaginas perveniant.* (Let a day be fixed for taking augury by the sacrifice of a dog before the corn comes out of the sheath and before it penetrates through into the sheath; i.e. booting stage = DC 45; Zadoks *et al.*, 1974).

[62] Red – Festus 45.10; reddish – Festus 397.25 (Lindsay, 1965). *Rutilae canes, id est non procul a rubro colore, immolantur, ut ait Ateius Capito, canario sacrificio pro frugibus deprecandae saevitiae causa sideris caniculae.* (Reddish dogs, that is not far from the red colour, are sacrificed, alleges Ateius Capito, to avert the Dogstar's rage by sacrificing a dog on behalf of the crops).
Elsewhere a suckling whelp is mentioned. *Fasti*: X-342ff.

[63] Biologically the passage makes sense. During a wet period with lodging of the wheat crop various fungal diseases get the opportunity to infect stems, leaves and ears. When the wetness is over and the crop dries the crop will look dirty, as if scorched. During good weather after a wet period dews may be heavy due to nightly distillation of water from warm soil to cool leaves. Such dews favour rust, as may be seen today on irrigated fields in a semi-arid environment.

[64] This is the time when the Scirocco may blow, a hot and dusty southern wind straight from the Sahara. People may suffer of drought, heat and dust ('scirocco' in Wikipedia, Italian version). In modern Rome, it is said, the frequency of suicides and car collisions increases markedly during Scirocco.

[65] Pliny *Naturalis Historia* XVIII: 284/7. Columella X: 341. ... *et tempestatem Tuscis avertere sacris*. (to avert by Tuscan [= Etrurian] rites the tempest).

[66] Chambry (1960: xiv) cited another 'international' tale, the story of the silver cup that Joseph placed in Benjamin's luggage (Genesis 44: 2), a trick also used against Aesop.

Chapter 2

[67] Just a few examples: encyclopaedia, general – Pancoucke (1777/1832, see under Tessier *et al.*, 1818), professional – Rozier (1809), mycology – Persoon (1801), De Candolle (1807, 1815), botany – Bulliard (1783), agriculture – Darwin (1800), Ponse (1810), Rozier (1809).

[68] Desmazières (1812: 122/3): '*La rouille est une poussière jaune, couleur de rouille de fer, que l'on remarque sur les tiges et les feuilles d'un grand nombre de végétaux, et particulièrement, dès le mois d'Avril, sur celles du blé. Cette poussière y forme des taches linéaires et parallèles;*' and '*... vraies plantes intestines, analogues à celles de la carie, et des genres* Urédo *et* Puccinie.'

[69] Seringe (1818: 201/2). '*... pustules ovales, extraordinairement petites, mais ordinairement très-nombreuses .... Et laisse voir une poussière jaune*'. '*...l'épuisement qu'avait produit la rouille...*'. '*Cette poussière ... présente ... des capsules ovoïdes, presque sphériques ...*'

[70] De Candolle (1815: 84 #624) came back from his original – and correct - opinion (Decandolle, 1807: 73) that *Uredo linearis* and *Puccinia graminum* were stages of the same fungal species.

[71] The color indications in the European trivial names of the cereal rusts date from Eriksson & Henning (1896). The 'color names' are used in all West-European languages. They gained official status at the various European Rust Conferences, among them the First European Yellow Rust Conference in Brunswick (Germany, 1956), the Colloque Européen sur la rouille noire des céréales (*[first] European Colloquium on black rust of cereals*) in Versailles (France, 1958), and the First European Brown Rust Conference, part of the Cereal Rust Conferences (plural!), at Cambridge (UK, 1964; proceedings published in 1966).

[72] De Bary (1866, 1867) demonstrated experimentally the existence of 'dimorphism', different spore forms belonging to one and the same rust species. Tulasne (1854: 81) discussed 'dimorphism' at length, with the side remark 'These opinions agree with the feeling of the farmers who believe that the orange *rusts* and the black *rusts* of the grain crops be but the different ages of a single parasite (F: *Ces opinions s'accordent avec le sentiment des agriculteurs qui veulent que les* rouilles *oranges et les* rouilles *noires ne soient que des ages différents d'un seul parasite*).

[73] *P. graminis* Pers. in De Candolle (1815: 59 #596) and *Uredo linearis* Pers. (p84, #624). Berkeley (1860 §601: 95), in accordance with some French mycologists (*i.a.* Decandolle, 1807), was convinced that *Uredo linearis* as the younger stage gradually changed into *Puccinia graminis* as the older stage (Berkeley, 1854/7).

[74] The sample consists of a segment of a wheat stem, the leaf and leaf sheath with about 50 per cent infection by black rust. The authors did show nor mention the glumes.

[75] '*Sur les chaumes, les gaines et les feuilles de plusieurs espèces de graminées*'. (On the stems, the sheaths and the leaves of several species of gramineous plants). The sample consists of a segment of a grassy

stem, possibly of couch grass, the leaves and leaf sheaths with about 50 per cent infection by black rust. Again, the authors did show nor mention the glumes.

[76] De Bary (1853: 102) mentioned the red-yellow uredial stages of *Trichobasis Rubigo vera* (DC.) and *T. linearis* (= present *P. graminis*) on stems and leaves and the red-yellow flecks on the beards and glumes of *Uredo glumarum* (*Trichobasis glumarum* Lév.). In the latter case, considering the confusing colour description, the plant part rather than the rust species dictated the rust's name. Both yellow rust and black rust can infect the awns, at least of wheat. Brown rust is not normally seen on glumes or awns.

[77] Westendorp (1860: 9 #25): *Puccinia recondita* Rob. – Desmaz. Pl. Crypt. De Fr., nouvelle série, no. 252: *Sur les feuilles languissantes du seigle*. (On the wilting leaves of rye).

[78] On rye – *Puccinia recondita* Roberge ex Desmazières; on wheat – *P. triticina* Eriksson.

[79] Among others Fuckel (1869) and Delacroix & Maublanc (1916: 156).
The confusion continued for a long time, as with von Thümen (1886 p20) who described the 'straw and stripe rust' of cereals (no crop specified) which occurs on culms, leaves and leaf sheaths. His *Uredo Rubigo vera De C.* seems to be a mixture of the brown and yellow rusts. It could appear in epidemic form on young wheat in the fall, sometimes so severe that fields had to be ploughed (probably yellow rust).
Pinckert (1867: 155/6) mentioned rust on rye. One he called 'crown rust' and the other 'grass rust'; the latter 'is more yellowish coloured and forms [?] stripes that follow the leaf veins ...'. The latter one was probably yellow rust. 'Crown rust' is a misnomer. *P. coronata* does not occur on rye. The original error was made by Kühn (1858: 104). 'Grass rust' (his '*P. graminis*') seems to have been used in a general sense. '*Die andere Form ist mehr gelblich gefärbt und bildet keimförmige* [?], *den Blattnerven folgende Streifen; man nennt sie Grasrost (*Puccinia graminis*).*' (The other form is more yellowish and produces germ-shaped [?] stripes which follow the leaf veins; she is called grass rust (*Puccinia graminis*)).
The apogee of confusion is in Karsten (1879: 29) who described the cereal brown rusts under the name *P. striaeformis* West. (syn. *Uredo rubigo-vera* De C.).

[80] Practice oriented Kühn (1858: 101) alluded to the overwintering of rust in the field but he did not see it for himself. He confused different rust species.

[81] De Candolle (1815: 83, #623d). This rust '*naît sur la surface supérieure des feuilles, et plus rarement sur la face inférieure, sur la gaine des feuilles ou sur la tige des graminées, et principalement du froment; elle y forme des pustules ovales extraordinairement petites, mais ordinairement très-nombreuses*'. The epidermis splits and '*laisse voir une poussière jaune: enfin, cette poussière devient rousse, mais jamais noire*'. '*... des capsules [= spores] ovoïdes presque sphériques ...*'.

[82] In this case brown rust of wheat, *Puccinia triticina* Eriksson & Henning.

[83] Lind (1913: 305) on *P. rubigo vera* de Candolle: 'From the accounts and descriptions of the more ancient authors it is sometimes to be perceived which species they have been dealing with, but as a rule they have dealt with all promiscuously.' On page 308 #1425 we find *P. glumarum* (Schmidt) Eriksson & Henning 1896, found on rye, wheat and grasses.

[84] In old documents there is an endless variation in the spelling of rust names.

[85] Persoon (1801: 216) wrote in Latin '*OBS. Vereor, [Vredo linearis] ne iunior plantula Pucciniae graminis modo sit.*'

[86] Fries (1829: 509) thought that from a taxonomic viewpoint *Uredines in Puccinias numquam abire possunt* (Uredos can never change into Puccinias) but added *Potius Uredo est Puccinia evolutione praecipita* (Rather the Uredo is a Puccinia with a rushed development). Apparently, Fries was somewhat in doubt.

[87] The colour change may have been from yellow to brown rust, or from the uredial to the telial stage. The latter option fits better with the remarks by De Candolle and others.

[88] Desmazière (1857: 798) had seen the brown rust of rye in the herbarium of Roberge, provided a good Latin description of the rust, and maintained its name as given by Roberge.

[89] In 1853 De Bary mentioned a glume rust (G: *Klappenrost*), *Uredo glumarum*, causing red-yellow flecks, possibly a composite of the yellow and black rusts, a confusing statement.

[90] Afraid to damage the specimen, I examined it by means of hand lens and torch. The rust I recognised as yellow rust, on the right hand leaf (upper side) with a severity of ~80 per cent. The leaves I believe to be rye leaves, one showing the adaxial side and one the abaxial side, but the identification is not definitive since auricles and ligulas were too much damaged. At places I saw the leaf hairs indicative of rye. Size and shape of the leaves were those of rye rather than of wheat, shown in a specimen with *Uredo linearis* (black rust). Leaf veins were somewhat less pronounced than in the wheat leaf, as normal for rye.

[91] Desmazières (1847: 10).

[92] Schmidt (1817) is said to have described yellow rust on glumes under the name of *Uredo glumarum*. The name was also used by Fries in 1821 (Eriksson & Henning, 1896). Westendorp (1854a) and Kickx (1867: 89) followed with *U. glumarum* Rob. ex Desm.

[93] Westendorp (1854a: 28, #135). *Uredo glumarum* Rob. – Desmaz. – HCB., no 568. The only comment was: *Dans l'intérieur des balles et des glumes du froment.* (Inside the chaff and the glumes of wheat). 'Inside the glumes' is typical for yellow rust. The critical Berkeley (see text) used Léveillé's generic name in *Trichobasis glumarum* Lév. (on glumes of cereals; Berkeley, 1860).

[94] Westendorp (1854a: 29, #144. *Uredo striaeformis* Nov. sp. '*Groupes linéaires, parallèles, allongés, nombreux, confluents, noirâtres, s'ouvrant suivant les sens des fibres de la feuille. Sporidies assez grosses, globuleuses, d'un brun foncé. .... Sur les jeunes pousses de l'Holcus lanatus et de l'Anthoxanthemum odoratum, dont il empêche le développement*'. (Groups linear, parallel, prolonged, numerous, confluent, blackish, opening in the direction of the veins of the leaf. Spores rather large, roundish, of a dark brown. ... On the young shoots of *H. lanatus* and of *A. odoratum*, of which it inhibits the development). The comment 'rather large' is meant in comparison with spores of stripe smut (*Ustilago longissima*) on *Glyceria* spp. Westendorp was followed by his fellow-countryman Kickx (1867: 55) with *P. striaeformis* West. (on straw of cereals).

[95] Kickx (1867: 55) quoting Westendorp gave a rather precise description of stripe rust, including the telial stage.

[96] Westendorp & Wallays (1855) N° 1077 '*sur les chaumes des céréales aux environs de Courtrai*'. A leaf fragment of ~10 cm long is shown, probably of wheat, not of *Holcus lanatus*. Telia are arranged in stripes. These telia are small, blackish, closed and shiny. They certainly do not belong to *P. triticina* or *P. graminis*. I did not see uredinia at macroscopic inspection.

[97] Fuckel (1869) mentioned telia and gave the yellow rust fungus the name *P. straminis*, which has no priority.

[98] Spahr Van der Hoek & Postma (1952, Vol. I: 624) told that rye in Friesland suffered a yield loss of ~50 per cent due to a kind of 'blast' (D: *brand*), that infected stems and ears and coloured them black (at least in wheat). The description reminds us first of black rust but it may also be inspired by the telia of yellow rust that can appear on stems and heads. 'Blast' was a frequently used but non-specific indication of diseases that coloured plant parts brown or black, as with rye in Roessingh & Schaars (1996: 110). Treviranus (1846) called the yellow rust 'glume blast'. Even the great de Bary (1853) used 'blast' (G: *Brand*) in a broad sense.

[99] Roessingh & Schaars (1996: 110). The text could be read as if 'honeydew' were not identical with 'red-dog' (1996: 434/5): 'Honeydew' on rye is, in today's view, a non-specific term then used for a cover on leaves due to either mildew, rust, the sugary excretion of aphids, or – on the ears – the sweet droplets produced by ergot.

[100] See also Bieleman (1987: 657) and note 585 (*ibid*.: 805). Tiesing (1923) mentioned 'red rust' (D: *roode roest*) on rye, probably meaning the brown rust (*P. recondita*).

[101] Now *Uromyces fabae* de Bary, the broad bean rust.

[102] Wiegmann (1839: 101): '*Der Honigtropfen war also offenbar zur Natur der Pilze übergegangen, seine Auszenfläche in die der Blattpilze, sein Inneres in die der Schimmelarten.*' (The droplet of honey apparently changed to the nature of a fungus, its outer surface in that of the foliar fungi, its inner side in that of the mould species).

[103] 'Philosopher' was a late 18th century word for 'scientist', student of the natural sciences.

[104] Fall infection of the recently emerged winter rye was not recorded in the Netherlands, 1845. Eriksson & Henning (1896: 146) mentioned severe fall infections of young rye in Denmark, 1874, and Sweden, 1887.

[105] My starting point is the 'General Report' (D: *Algemeen Verslag*) over the year 1846, written by a committee of the 'Netherlands Company for the Promotion of Industry' (D: *Nederlandsche Maatschappij ter Bevordering van Nijverheid*), seated at Haarlem. The Committee consisted of knowledgeable persons, working at the request of the Company's Board, who collected data from provincial informants.

[106] The decisive characteristic mentioned by Göppert (1846) is the 'skin' of the spores, consisting of 'a water-clear transparent skin' (... *einer wasserhellen durchsichtigen Haut* ...). Anonymous (1846b: 48) states 'dust ... consists of small round grainlets, yellow in color and translucent in the middle' (*stof ... bestaat uit kleine, ronde korreltjes, geel van kleur en in het midden doorschijnend*). See Cummins (1971: 151) – wall pale yellowish or nearly colorless; Savile (1984: 99) – wall hyaline (subhyaline). The other rye rusts have coloured urediniospore walls.

[107] In the duchy of Limburg (S Netherlands) night frost damaging the tips of the rye ears was mentioned explicitly (General Report 1846: 27)); same in Gelderland (Anonymous 1846b: 56).

[108] The names and places mentioned in this paragraph are shown on the map of the Netherlands around 1846 (Figure 2.6).

[109] With some good will one could think of telia of yellow rust, but all telia are black and not chocolate-coloured. The description does not fit the black rust.

[110] Banks was inclined to believe that the 'rust', the uredinial stage of black rust on wheat, and the 'blight', the telial stage, were two different species; Banks preferred to follow Fontana (1767).

[111] Van Hall (1859) followed and quoted Unger (1833) and in the postscript the learned Harting did not disagree.

[112] See congress reports in Anonymous (1846/7). The text paraphrases two passages: 1846 *'de bezwarende verschijnselen in de aren der rogge waargenomen, twee derden verloren. Uredo rubigo op de bladeren, geen vrees, gaat niet verder. U. caries of U. segetium op de schutblaadjes van de aartjes, daarna op korrel zelf, op vele akkers de oogst hebben weggenomen'* and 1847 *'en algemeen was men het daarover eens, dat het verwaaijen van het stuifmeel, dat het vorig jaar was achtergebleven, en thans in zoo ruime mate had plaats gehad, dat men somtijds in eene wolk daarvan gehuld was, op grond van ervaring het vaste vertrouwen gaf, dan men van de roest, al kwam ze nog, niets meer te vrezen had.'*

[113] In the 'General Report' on 1852 complaints about rust in rye were mentioned (Friesland, p83); the annually returning rust in rye did much damage (Groningen, p87). It is a fair guess to identify this rust as yellow rust.

[114] The names and places mentioned in this paragraph are shown on the map of Europe around 1846 (Figure 2.12).

[115] In this way the 'field races' of yellow rust on wheat were discovered. Later, their existence was demonstrated experimentally (Zadoks, 1961).

[116] Bjerkander (1794 Num. 98) *Secale cereale. Rågen sades vara denne Sommar ymnig öfver hela Riket. Härstädes hade han på några ställen mycken rost. Den 8 Junii syntes alla Rågbladen gulla, när de granskades med et stort Microscop, voro de beklädde liksom med små blemmor, hafvande uti sig et fint gult mjöl, som skådadt genom Microscopet, bestod af små gula Gerber. Den 6 Julii voro Axen gula af samma rost, somlige hade uti sig ingen Råg, utan voro aldeles fördärsvade, somlige ägde ej större än gryn och alla voro i topparna öfverklädde med de gula Gerberna eller rost.* (This summer the rye crops were abundant over the whole country. Nearby we have had much rust in some places. On 8 June all leaves were yellow, upon examination under a strong microscope they were covered as with small pustules, that contained a fine yellow powder, and seen under the microscope consisted of small yellow [powder]. On 6 July the ears were yellow by the same rust, and some had no kernels in them, but were completely destroyed, some were no larger than [grains], and all were covered with the yellow [powder] at the tips).

[117] Anonymous (1850b); Blok (1977: 59/91).

[118] Ludolph Christian Treviranus, professor of botany at Bonn University, must be considered a reliable observer. He wrote an impressive textbook *Physiology der Gewächse* (Plant physiology), 1835, 1838.

[119] Léveillé (1848: 777). *'Une espèce (*Uredo glumarum *Rob. in Dsmz., Pl. crypt. de France, éd. 2, no 107, 6; et An. Sc. Na., 3e sér., tom. VIII: 10), qui a beaucoup d'analogie avec la Rouille, s'observe sur les glumes du Froment et du Seigle qu'elle déforme, et dont elle produit quelquefois l'avortement.'* (A species (...), that shows much analogy with the Rust, is seen on the glumes of wheat and rye which she deforms, and of which she sometimes causes abortion.)

[120] Weber (1847: 5) ... *die grosze Strecken verwüstete* ...

[121] In 1853 De Bary still followed the customary separation between *Uredo* and *Puccinia*. He mentioned *Trichobasis Rubigo vera* (DC.) and *T. linearis*, yellow-brown rust and black rust, following Léveillé (1848). He also mentioned *U. glumarum* as glume rust (G: *Klappenrost*), without making clear which rust he meant.

[122] Léveillé (1848: 777) '*Dans une note que j'ai reçue de M. Auerswald, j'apprends qu'elle a été très funeste en Saxe il y a trois ans. Nefaria ista pestis anni 1846, telles sont les expressions dont il se sert pour me peindre ses effets.*' (From a note which I received from Mr. Auerswald I understood that she had been very disastrous in Saxony three years ago. This darned pest of the year 1846, that is the expression which he uses to paint me its effects).

[123] From Eriksson & Henning, 1896.

[124] E: Sambia, G: *Samland*, the peninsula NW of Kaliningrad. Von Viebahn (1848: 43). The disease could have been yellow rust, black rust or *Fusarium*.

[125] Data were derived from a well-designed enquiry among the Agricultural Societies of the Kingdom Prussia yielding a large but non-specified number of entries, with yield estimates weighted for areas covered. The 'norm' was not indicated; it probably was something like the average yield over 1840 to 1844. The Agricultural Societies represented the estate owners with relatively high standards of crop husbandry. The small farmers, often with low standards of crop husbandry, have most probably suffered larger losses. The impact of small farmer information on the tabulated data can only be guessed.

[126] Anonymous (1847b). '*... dasz die krankhafte Erscheinungen, welche früher nur auf den eigentlichen Blättern bemerkt worden, nunmehr auch auf die Blattscheiden, den Halm, selbst auf die Aehren und das Korn übergegangen, und dasz dieselben all überall zu finden seien: - keine Bodenart, keine Bestellungsweise, keine Vorfrucht scheine eine Ausnahme zu machen: je weiter der Roggen entwickelt, in einem desto höheren Stadio zeige sich auch bei ihm die Krankheit.*' The text continues with an utterly confused attempt at naming the rust.

[127] Windt (1806: 58), who worked in a similar area (near Minden), tentatively mentioned that the black rust appeared in rye around barberry shortly before or during flowering.

[128] Generally speaking this observation is correct but it is not specific for any rust species. Pliny (*Naturalis Historia* XVIII: 154), Theophrastus (VIII, x, 2; Hort, 1916 Vol. 2: 203), and Windt (1806: 23) made the same observation.

[129] These observations were confirmed in Austria, 1916, by Fruwirth (1916), who mentioned empty spikelets and empty ('deaf') ears of rye after a severe yellow rust attack on the glumes.

[130] In wheat I did not see such inhibition of fructification. When the yellow rust infection was severe the grain setting usually was regular but the grain quality might be quite poor, light weighted, 'chicken fodder', as the farmers said. Infertility did sometimes occur at the tip of the ears.

[131] Regel (1854) '*So der in Form von gelbbraunen Punkten und Streifen am Getreide erscheinende Rost (*Uredo Rubigo vera *und* linearis). *Derselbe ist in manchen Jahren sehr häufig, bedingt das sogenannte Vergelben des Getreides und leichte Frucht. So richtete er z.B. 1846 vielen Schaden bei uns an.*' U. Rubigo-vera is the

yellow rust, *U. linearis* the black rust. Apparently, the yellow rust dominated in 1846. NB. Barberry and black rust used to be common high up in the Alpine valleys until cultivation of the Alpine fields lost its profitability after World War II.

[132] This passage attempts a reconstruction of past conditions favouring a yellow rust epidemic based upon recent knowledge (Zadoks, 1961) and scanty information from 1845/6. More detailed weather data from 1845/6 were available (Bourke & Lamb, 1993), but I don't believe that their analysis would contribute essentially new information.

[133] Vanderplank (1968: 155ff). The selection pressure by tropical maize rust (*Puccinia polysora* Underw.) on maize, a cross-fertiliser as rye, in W Africa lasted for years, allowing gradual genetic adaptation of the crop toward higher resistance levels by the accumulation of minor genes. It is difficult to imagine that a strong but once-only selection pressure, as in 1846, could have a similar effect.

[134] Appel (1931), working in Germany, mentioned the existence of yellow rust on rye in passing, without epidemiological details.

[135] Shrivelled grain is in the handbooks, as in Eriksson & Henning (1896). It can also be caused by black rust (Lehmann *et al.*, 1937: 372), more rarely by brown rust.

[136] This epidemic is not listed in Zadoks (1961) or Hogg *et al.* (1969). Around 1825 the sensitivity of rye to inundation (winter rye), drought, summer rainfall, and 'red dog' (D: *roodhond*), possibly brown leaf rust (*Puccinia recondita*), was well-known (Roessingh & Schaars, 1996: 110). Kühn (1858: 101) may have suspected the overwintering of rust on rye, by mentioning fall infection on early sown crops and early May appearance; he described *Uredo Rubigo-vera*, merging the present *Puccinia triticina* (on wheat), *P. recondita* (on rye) and *P. coronata* (on oats). Rust on rye could at times cause important damages (*ibid.*: 100), he wrote. Which rust?

[137] Contemporaneous scientific papers on the 'potato murrain' abounded (Chapter 3) but those on the rust of rye were scarce indeed. The catastrophe of the potatoes by far overshadowed the failure of the rye. The excellent agricultural history of the Netherlands edited by Sneller (1943) discussed the 'potato murrain' seven times but the rust on rye was never mentioned.

[138] This I observed during the hot and dry summer of 1959 (Zadoks, 1961: §37.2).

[139] Bjerkander (1794).

[140] See Chapter 5.

[141] Denmark (Kølpin Ravn, 1914: 112), Sweden (Eriksson & Henning, 1896: 203).

[142] Around the mid-19th century cereals on the löss soils of Limburg covered about ¾ of the arable land, with rye as the main crop (Bieleman, 1992: 159/60). On the sandy soils crop rotation was restricted, with much rye (van Zanden, 1985: 168), which may have facilitated the switchover of the rust.

[143] Ponse (1827) began with an excerpt from Banks (1805). The description clearly refers to rust on wheat. Ponse's rust is a combination of yellow rust (early in the season) and brown rust (late in the season).

[144] In the 17th century and long after, apparently up into the mid 19th century, much rye was grown on a particular sandy soil type (D: *geestgronden*) behind the dunes of N and S Holland (Bieleman, 1992: 75).

[145] At the time, rye was also grown on clay soils; rust severities on clay and sandy soils were about the same.

[146] During rain showers I used a suction device, placed in a crop severely diseased by yellow rust, to collect surface water from healthy plants. The water from the healthy leaves was heavily loaded with urediniospores of yellow rust.

[147] 'The red' (D: *Het roode*) is a term for rust (Roessingh & Schaars, 1996: 446) and probably 'red' (D: *rood*) also.

[148] The unusual text from Bavaria reads as: 'In winter rye the yellow blast (G: *Brand*) was seen at many locations. Yellow exudation at the tip of the kernel. Each glume, where the curling at the tip of the kernel was yellow, was also covered with yellow dust. Such diseased kernels were sweeter than the healthy ones. Each ear had two or more yellow-blasted kernels.' *Author's comment*: The sweetness reminds of ergot rather than yellow rust but the yellow powder points to yellow rust. No mention was made of ergot at harvest. Yields were reported to be good. So, what caused the affliction?

[149] Oekonomische Neuigkeiten und Verhandlungen 73 (1847): 199.

Chapter 3

[150] For the new taxonomic position of *P. infestans* see Govers & Gijzen (2006).

[151] The original North American and Irish (= European) strain(s) of *P. infestans*, closely related, were from Andean origin (Gómez-Alpizar *et al.*, 2007).

[152] Bourke & Lamb (1993: 49).

[153] Bourke (1993: 12) – Ireland, 4.5-6.3 kg/day, 5.4 on average; Connell (1950: 149) – on average 4.5 kg/day. Hooijer (1847: 1, on the Bommelerwaard, NL): 'The poorest ate potatoes only, cooked at noon, as porridge at supper, fried at breakfast', no amounts specified. Note: An active young male needs ~3000 kcal/day equivalent to roughly 3.75 kg of potatoes per day; a labourer needs more for heavy manual labour. Note that a diet of potatoes only, supplemented by some fresh milk or buttermilk, was healthy and allowed to raise large families, according to Reader (2008). Curiously, Van Bavegem (medical practitioner at Dendermonde, Belgium, between Gent and Antwerp) mentioned the same diet for poor families in 1782 (Van Bavegem, 1782: 28).

[154] At the time, 1845/50, the differences between typhus, typhoid and several other diseases were not so well defined as today. Whereas 'hygienists' (today we would say 'epidemiologists') had fair ideas about the environmental conditions favouring one or the other disease, the causal agents and the modes of transmission of these epidemic diseases were unknown. We distinguish typhus, caused by a rickettsia (*Rickettsia prowazeki*) and transmitted by body lice, and typhoid, caused by a bacterium (*Salmonella typhi*) and transmitted by water. The accurate description by Virchow (1848) clearly points to an epidemic of typhus. In view of the hygiene, or better the lack of hygiene, on the countryside we might think of typhoid when the original texts are not clear.

[155] Bourke, 1993; Dyson & Ó Gráda, 2002; and many others.

[156] Dupâquier, 1980: 173; The 'Census Commission' of 1851 said 'very nearly one million' (total or excess) deaths? (Woodward, 1938: 340). Woodham-Smith (1962: 411) stated that the Irish population dropped between 1841 and 1851 from 8,175,124 to 6,552,385, a loss of 1,622,739 persons, adding that the population censuses were quite unreliable.

[156a] An abridged version of this chapter was published as Zadoks (2008).

[157] Good data are scarce. Geertsema (1868: 218) quoted data for a farm family with 3 children and 3 personnel that suggest a daily intake of ~2 kg/day per adult; Priester (1991: 365) quotes an estimate from 1882 for a labourer's family, man, woman, 3 or 4 children of 1,600 l/year, or ~3 kg/day per family, or probably well over 1 kg/day for the labourer; in either case I suppose that the potatoes were supplemented with other food.

[158] Wehnelt (1943: 178) quotes Dietrich (1850) with a similar clause by a pastor in Giebichenstein near Halle (Germany). The interesting detail is the reference to the sloppy cultivation of potatoes for distillation purposes.

[159] Brabant for short, officially Noord Brabant.

[160] Belgium – Van Bavegem (1782); Belgium and the Netherlands - Van der Zaag (1999: 93/5); England – Vanderplank (1949); France – Van Aelbroeck (1830: 187); contemporaneous instructions – Soetens (1834), Albert (1845), Uilkens (1853: 43).

[161] Potato virus Y and several other potato viruses are transmitted by aphids, which are less frequent in cool, moist and windy areas.

[162] Salaman (1949: 120); Van der Zaag (1999: 92/3); contemporaneous recommendation – Uilkens (1855: 87).

[163] Green lifting implies potatoes being lifted (harvested) when the foliage is still green and lush, in the expectation that eventual viruses have not yet descended from the foliage into the tubers. Before green lifting the foliage must be destroyed.

[164] Friesland – 1807 ex Spahr Van der Hoek (1952: 646), 1809 ex Kops (1810: 357); Scotland – ex Van der Zaag (1999: 92).

[165] A few authors stated that the late blight disease was first found in Denmark, 1842 (De Bary, 1861; Gram *et al.*, 1940: 68). This date is not confirmed by the overwhelming evidence presented in this chapter. I presume that the 1842 quotation is due to an erroneous interpretation of a report (not known to me) on the hard rot disease of potato. In the early days the two diseases, due to a *Fusarium* and a *Phytophthora*, respectively, were often confounded.

[166] According to Bourson (1845) my statement would be too strong. On 18 July, 1843, the Provincial Council of West Flanders voted 2000 BFR to import potatoes from the USA. Small quantities may have been imported but the government refused to import potatoes by the shipload (F: *cargaison*). Though foreign potatoes had been tested at Cureghem already for at least two years, Bourson is not explicit as to the actual planting of tubers with USA provenance. He mentioned that at least one private person

experimented with potatoes obtained from the USA [probably at the University of Gent]. Bourson's vagueness may have been deliberate as he worked for the Belgian Ministry of the Interior.
I could not locate Cureghem in W Flanders. There is another Cureghem, a district in the municipality of Anderlecht, in the W of the Brussels agglomeration.

167 Great Hunger - Woodham-Smith (1962); Black Years – Hoogerhuis (2003: 31).

168 I venture two possible, complemenary explanations, inadequate ripening of seed potatoes lifted in a wet autumn 1845 and stored at relatively high temperatures, and drought combined with high temperatures during early summer 1846.

169 The observation that late blight in England began in low places was repeated many times in The Gardeners' Chronicle, 1845, as on page 592.

170 Decaisne (1846) quoted extensively from Desmazières; Bourson (1845: 2698); Roze (1898: 292).

171 Lestiboudois (1845: 245).

172 Vis (1845: 2-3) stated that the favoured variety Beaulieu was lost in 1844 'by the blackening of the leaves and the dying of the plant' (*door zwart worden der bladeren en versterving der plant*). He added that the disease could not have come from the roots as in that case the foliage would have yellowed first.

173 Colza, swede rape, rape seed, *Brassica napus* L.

174 The Netherlands – Van der Zaag (1999: 108); general – Abel (1974: 365).

175 Un-harvested potatoes producing volunteer plants in the following season.

176 Heesterbeek & Zadoks (1987); Zadoks & Van den Bosch (1994). In contemporaneous language: *la marche de la maladie était rayonante et successive* (the course of the malady was radiating and gradual; Bourson, 1845: 2720).

177 Two contemporaneous references found: Sauberg (1845: 30) 'The last, worst and by wet rot affected potatoes remain lying on the field' (G: *Die letzten, schlechtesten und nasse Fäulnis übergegangenen Kartoffeln bleiben auf dem Felde liegen.*) Anonymous (1846d: 213), in Overijssel, Netherlands: 'In the spring [of 1846] self-sown potatoes also appeared everywhere, that had been left on the field during reaping or had been thrown aside as diseased or decayed, whereas the frost had not killed them during the mild winter [1845/6].' (*Eveneens sloegen in het voorjaar overal de aardappels op, welke op den akker bij het rooijen achtergelaten of als ziek en bedorven weggeworpen waren, terwijl gedurende den zachten winter de vorst deze niet had gedood*). Note that Fürnrohr (1845) recommended removal of rotting tubers from the field.

178 Among others *Bellis perennis, Primula auricularia, Senecio vulgaris, Viola tricolor*; mean temperature in January 1846 was 5.1 °C, far beyond the 1833-1845 average of 1.9 °C.

179 *Allgemeine Zeitung* #214, August 2nd, p1708. *Die andauernde Hitze schmeltzt überall das Eis auf den Bergen. Die Spitze des Montblanc ist gegenwärtig nackter Felsen, seit vielen Jahren war das Eis dort nicht verschwunden. Mehrere Flüsse sind in Folgedessen aus ihren Ufern getreten, so der Rhone, welcher im Kanton*

Wallis neuerdings 1000 Juchart Acker überschwemmte. The Mont Blanc (4807 m) is the highest mountain of the Alps. The *Allgemeine Zeitung* was the semi-official organ of the Austrian government (Blum, 1948).

[180] Allgemeine Zeitung #194 July 13th p1549 (Bavaria); #198 July 17th p1583 – a destructive hurricane swept over Bohemia; #217 August 5th p1723 – a severe storm broke or even uprooted many trees and hail destroyed the crops near Füssen in S Bavaria; etcetera.

[181] Strictly speaking these indexes, developed by IJnsen (1976 and 1981), are valid only for De Bilt in the centre of the Netherlands, but they are indicative for the Netherlands and even beyond (the West European Climate Zone). High values indicate hot summers and cold winters, respectively. The indexes run from 0 to 100. The summer index has a normal distribution. The winter index is skewed to the right.

[182] In his diary Hoekstra (1879) mentioned 16 March, 1845, as the coldest day since time immemorial. On 25 March harness races for horses had been organised on the ice of the Zuyderzee near Lemmer. At the time the Zuyderzee was a salt water bay of the North Sea. Records of several similar exploits exist.

[183] Hoekstra (1879) in his diary called the summer of 1846 the hottest summer within living memory. In fact, it was the second hottest summer of the 19th century, the hottest being 1868.

[184] Ising (1892: 71) *Alle andere kunstmatige middelen liggen, naar onze overtuiging, buiten den kring van pligten der Regering.*

[185] The ~60 per cent loss in 1845 can be ascribed do the direct effect of late blight on the potato crop. The ~40 per cent loss of 1846 is largely due to indirect effects. (1) A reduced potato area. (2) Transfer of potato cultivation from clay soils to less productive sandy soils. (3) Shift to less productive early potatoes. (4) Early summer drought. Direct effects occurred in places, usually less than in 1845.

[186] Groningen (Priester, 1991: 342); Beijerland (Baars, 1973: 136), text of page 136, not clearly confirmed by Fig. 61 on page 218; Bommelerwaard (Bieleman, 1992: 152); etc.

[187] Bieleman (1987: 550); Priester (1991: 342). Generally speaking, potato crops on the sandy soils of the Netherlands have a dryer micro-climate than on the clay soils, less conducive to late blight.

[188] At the time, the Austrian-Hungarian double monarchy with the Austrian Crownlands covered a large part of Central Europe, stretching roughly from S-Poland to N-Italy, and from Switzerland to the Ukraine. In this paper Austria encompasses Imperial Austria and its Crownlands, with ~25 million inhabitants (Bolognese-Leuchtenmuller, 1978; Hungarian Kingdom excluded).

[189] Hlubek (1847: 65), Unger (1847: 307).

[190] Macartney (1968: 313), Blum (1948: 110).

[191] Lamberty (1949: 136); Lindemans (1952: 188).

[192] See also *Staats-Courant* 1845 N° 237 dated 7 October. At the time the Dutchies Schleswig and Holstein were under Danish rule. As Schleswig-Holstein they presently are a state of Germany, except for a northern strip that voted to remain Danish.

[193] Lille – Lestiboudois (1845: 245). SW France (Aquitaine) – Armengaud (1961: 173). Duchy of Savoy – Bonjean's (1846: 122) calculation of the loss in Savoy has to be updated. If a is the (known) amount harvested and b is the (estimated) amount lost, Bonjean calculated the loss in terms of percentage as 100 x b/a = 100 x 1,645,976 / 2,002,914 = 82.2 per cent. Today we prefer to calculate the loss as 100 x $b/(a+b)$ = 100 x 1,645,976 / 3,648,890 = 45.1 per cent. Bonjean's (*ibid.*: 123) estimate of financial loss is different. He valued the harvest at 15.5 million francs (on a shaky base) and estimated the loss at 5 million francs (without underlying data). At constant prices this would mean a loss of 24.4 per cent. We reject Bonjean's original figure (86 per cent) and prefer the use Bonjean's tabulated data leading to ~45 per cent loss.

[194] This term was borrowed from Newman & Simpson (1987) to indicate the various countries forming the German Federation (G: *Deutsche Bund*) and the areas under their rule. Silesia (in present SW Poland), for example, though ruled by Prussia, the most powerful country of the German Federation, did not belong to the Federation. Austria belonged to the German Federation but not the many areas ruled by Austria, Hungary foremost. Austria is discussed separately.

[195] The combination of light foliar infection and considerable tuber infection is not exceptional, occurring when the rains set in after a drought period, maybe because spores from the foliage are washed down to the tubers through cracks in the soil. It happened in 2006 in some areas in the Netherlands (Besteman, 2007: 209).

[196] Weber (1847) reported on the Germanies, 1846: Brunswick – Potatoes healthy but loss (by drought?) 25 per cent; Poznań – sometimes no tubers, only long roots; Prussia – in some areas loss (by drought) over 60 per cent; Rhine Province – losses due to late blight; Silesia – In Upper-Silesia fields had to be ploughed in. In Middle-Silesia poor tuber setting, much secondary growth after July rains, and late blight. In Lower-Silesia drought damage.

[197] Anonymous (1848b: 373): potato losses in Kingdom Prussia 31 per cent of average, in Silesia 61 per cent of average; Virchow (1848: 37) in Upper Silesa nearly 100 per cent.

[198] See also *Staats-Courant* 1846 N° 21 dated 24 January: 2.

[199] Oekonomische Neuigkeiten 71: 204 (G: *Die Noth is deshalb unbeschreiblich grosz*).

[200] At the time Switzerland was a geographic area, a conglomerate of fairly independent cantons, rather than a nation-state. After the civil war fought in 1847 it became a nation-state.

[201] The comments are quite pertinent in Quanjer *et al.* (1916: 12) and Quanjer *et al.* (1920: 54). In the latter paper Dorst refers to a report of the 'Royal Agricultural Society', Londen, 1872, stating that many potato varieties were degenerated and consequently highly susceptible to late blight. This interaction was mentioned again by Schick & Klinkowski (1962: 1175, interaction with potato viruses X and Y).

[202] Allgemeine Zeitung 1846 #198 July 17th: 1583.

[203] Note that Prussia is the name of a kingdom and of a province or area within that kingdom.

[204] Mecklenburg – Pogge (1893: 58); Silesia – Oekonomische Neuigkeiten und Verhandlungen 71 (1845: 64, 191).

[205] Anonymous (1846a: 193).

Notes

206 Jansma & Schroor (1987: 59); Bieleman (1992: 51).

207 Allgemeine Zeitung 1846 #198 July 17th: 1583.

208 Pirenne (1932: 128), Dumont (1977: 399).

209 Agulhon et al. (1976: 140), Jardin & Tudesq (1973: 234), Le Roy Ladurie (2004: 623), Newman & Simpson (1987: 383).

210 See Table 2.1; Wehler, (1987: 643); several authors reported severe rust in wheat as in Brandenburg and West Prussia.

211 Von Viebahn (1848: 14) visited the district along the river Netz (presently N Poland) on 10 June, 1846: 'Der Stand der Früchte in dieser Landschaft hatte durch die anhaltende Dürre wesentich gelitten; insbesondere der Roggen stand an manchen Stellen so slecht, dasz keine ergiebige Erndte mehr gehofft werden konnte.' (The stand of the crops in this area suffered materially from the continuing drought; the rye especially was so bad in many places that a good yield could no longer be hoped for).

212 W Prussia – Anonymous, 1846a; Bavaria – Allgemeine Zeitung 1846 #194 July 13th: 1549.

213 Allgemeine Zeitung 1846 #212 July 31st: 1694.

214 Wallis – Allgemeine Zeitung #201 July 20th: 1607; Lichtenstein – Allgemeine Zeitung #202 July 21st: 1612.

215 Austria – Macartney (1968: 313); Silesia (Prussia) – Anonymous (1848b: 381). Wikipedia – famine general.

216 Experimental techniques were yet in the suckling stage. Fungal spores will infect plants only when the environmental conditions are just right. If not, fungal spores will not germinate and penetrate their host plant. Then, the serious experimentalist is bound to take sides with the anti-fungalists (e.g. Decaisne, 1846: 12; Gadichaud, 1846: 281; Harting, 1846: 247; Martens, 1845: 365/6; Unger, 1847: 309/10; Schacht, 1856: 9). According to my experience experimenters must have the right mind-set to persevere and bring their experiments to a success. Maybe the anti-fungalists, 'exanthemists' and 'environmentalists', had a prejudice against the concept of contagiousness of fungal disease and so did not persevere in their efforts.

217 Gebel in Annalen der Landwirthschaft 1848, Vol. 11: 156 'das geschwächte innere Leben der Kartoffel'. Nitsch (1846: 291) 'Unläugbar ist es, dasz die Kartoffelkrankheiten in der gesunkenen Lebensenergie der Pflanze liegen und dadurch ein krankhafter Vegetationsprocesz bedingt ist.' (It is undeniable that the potato diseases lie in the sunken life energy of the plants and that this determines a sickly vegetation process). Note: Such views are typically echoes of 'romantic phytopathology' (Wehnelt, 1943).

218 Van Peyma (1845: 20) 'The contamination and infection then occurs as it were quick as lightning through air and wind stream, ...' (De besmetting en aansteking geschiedt dan als het ware vliegend snel door lucht en windstroom, ...); Committee (1845), the Committee described an incredibly fast infection starting from initial foci, mentioning that many had seen a downwind dispersal, and delayed infection of fields sheltered from the wind by forest.

[219] In his review the Dutchman Uilkens (1852: 77/9) accepted the infectious nature of the fungus and the wind-borne character of the epidemic.

[220] A similar argument was worded in France (Lestiboudois, 1845: 263) '... by the noxious action of a miasma among plants, the application of that doctrine on plants will be pushed back by the facts ...(... *par l'action délétère d'un miasme chez les végétaux, l'application de cette doctrine à la maladie actuelle serait repoussée par les faits* ...) and on p263 he quoted the otherwise well-informed Desmazières '... because, according to him, it is unheard that a cryptogam would completely destroy a harvest, ... (...*parce que, suivant lui, il est inouï qu'une cryptogame détruise complètement une récolte,* ...).

[221] Among them Harting (1846: 288). *La disposition à être atteint par le mal n'a pas été la même pour toutes les variétés des pommes de terre* (The disposition to be affected by the blight has not been the same for all potato varieties); Martens (1845: 359) thought of quality differences among seed potatoes; see also Hlubek (1847: 65); Lestiboudois (1845: 247); many others.

[222] Committee (1845) *Het loof der zieke aardappelen, dat buitendien geene waarde heeft, worde op het veld verbrand en men verniele tevens alle onbruikbare en verrotte aardappelen, zoodat er van het ziekelijk gewas zoo min mogelijk op het veld overblijve.*

[223] Foliage (Morren, 1845a), tubers (Hlubek, 1847: 69).

[224] Diluted sulphuric acid: Moleschott & Von Baumhauer (1845: 15); Morren (1845a).
Chalking: Molenschott & Von Baumhauer (1845: 1518); Morren (1845a). The latter advised to use a mixture of chalk, salt and copper sulphate, in water or as a dry powder.

[225] Morren (1845b: 376). '*Toutes les récoltes de pommes de terre faites autour des usines de zinc, à Angleur, à St-Léonard, à la Vieille-Montagne, dans le cercle d'action des substances volatiles qui s'échappent autour de ses usines et qui font tant de ravage parmi quelques espèces d'arbres, ont été excellentes et à l'abri complet du fléau.'* '... *et ce fait est de la plus haute importance, car il ne peut s'expliquer que par l'action d'une substance métallique comme matière de chaulage sur la végétation.*' (All the potato harvests made around the zinc factories at Angleur, St-Léonard and Vieille-Montagne, within the action radius of the volatile substances that escape around their factories and that do so much damage to some tree species, have been excellent and completely protected from the scourge. ... and this fact is of the highest importance, because it can be explained only by the action of a metallic substance as a chalking [= chemical treatment] on the vegetation).

[226] A similar observation was made on potatoes growing around copper smelting works in Wales, 4 September 1846 (Reader, 2008: 206).

[227] Bergsma (1845: 9): '*Niet zelden zag men een land met aardappelen in een paar dagen geheel van voorkomen veranderen en bespeurde men, vooral des avonds, eenen ondragelijken stank, welke zich op eenen aamerkelijken afstand verspreidde'. (ibid.: 30): 'De waarnemingen dat de ziekte bij derzelver verspreiding dikwijls de rigting van den wind gevolgd is, wordt nog meer waarschijnlijk, daar sommige, achter heggen of boomen groeijende aardappelen, verschoond gebleven zijn, en eerst later de ziekte gekregen hebben.' (ibid.: 15): '..., dat wij tot het besluit moeten komen, dat de krul, roest en kankerachtige ziekte van elkander niet verschillen.'* Bergsma was deeply impressed by the potato dry rot epidemic (Von Martius, 1842) that had already affected many Dutch potato fields. He had translated Von Martius' treatise into Dutch. The word 'rust' (D: *roest*) for potato blight came out of the blue. Bergsma apparently used it in a non-specific way similar to the old usage of the word 'blight'.

[228] The stench was a typical symptom at field level, mentioned by several authors (e.g. Bergsma, 1845); very unlike the smell of a healthy potato field (Schacht, 1856).

[229] Harting (1846), an influential biologist, performed quite sensible infection experiments that, unfortunately, failed. Thus he became an anti-fungalist, referring to Unger (1833).

[230] Vrolik *et al.* (1846: 12) *Dat de proeven, hoewel, wat de ziekte betreft, niet aan het oogmerk beantwoord hebbende, echter, door het op zeer onderscheiden gronden voortbrengen van nieuwe soorten ... niet onbelangrijk geweest zijn, in vele opzichten leerzaam en niet geheel ongunstige resultaten voor de toekomst belovende.*

[231] Butler & Jones (1955: 521); Uilkens (1855: 39); Van der Zaag (1999); Vanderplank (1968: 153) provided a 20th century empirical argument.

[232] Enklaar (1846: 273) stated that the more tasty varieties suffered more from the disease.

[233] Desmazières described *Botrytis fallax* in 1844 already but he rather thought it to be the consequence and not the cause of the disease. The priest Van den Hecke identified a *Botrytis*, 31 July 1845. A fungus was inculpated by Martens on 14 August, by Vanoye and by Morren on 18 August 1845 (Bourson, 1845; Semal, 1995).

[234] Fürnrohr (1845): Selection of healthy tubers for planting in 1846 and continuous health surveillance, winter storage of consumption potatoes, industrial use of diseased tubers, removal of rotting tubers from the field.

[235] Unger (1833) claimed that fungi, that we now see as pathogens, were not the cause but the consequence of disease; they appeared rather as a kind of 'rash' (*exanthema*) on internally diseased plants.

[236] Sneller, 1943: 73ff, 434; Van Zanden, 1985: 116 (*starheid van het loonpeil* = 'rigidity of the wage level', specifically in the first half of the 19th century).

[237] Around 1850, a landless labourer rented a potato plot of about 1/3 ha (with hog ½ ha) at the price of twenty days' work (Van der Zaag, 1999: 111).

[238] In his eulogy on the potato Parmentier (1781: 151) already quoted the beneficial effect of potato eating on sailors suffering from scurvy.

[239] E.g. Agulhon *et al.* (1976: 10); Hofstee (1978); Figure 3.7.

[240] Augustin Sageret in Tessier *et al.* (1818: 756) wrote enthousiastically that one *arpent* (an area measure) of potatoes could feed 5-6 times as many persons as one *arpent* of wheat. A more normal ratio was 2 to 3 times. Slicher Van Bath (1987: 293 and 407 – note 205) and Reader (2008: 128) quoted authors relating rapid population growth to the then new cultivation of the potato. For a similar, contemporaneous remark see Hlubek (1847: 65).

[241] Many authors, e.g. Bergman (1967: 400); Hofstee (1978); Lutz (1985: 244). In Drenthe (NL), a poor province, the birth rate decreased from ~10 (mean over 1841/45) to 3 per cent (mean over 1846/50; Bieleman, 1987: 97). Zeeland – Priester (1998: 77).

242 500,000 (Connell, 1962: 64; Dupâquier, 1980), 2 million (Woodham-Smith, 1962).

243 Hofstee (1978: 212). Excess deaths not due to cholera amounted to ~53,000, disregarding the provinces of Brabant and Limburg, over the years 1846-1849. This figure approaches the 60,000 of Turner (2006: 342, no source given).

244 The late J.B. Ritzema van Ikema, when president of the Wageningen Agricultural University, used to tell me about letters written by one of his forefathers, a well-to-do farmer in the NW corner of Groningen, neighbouring the sea, complaining about the poor quality of the quinine delivered by the Groningen chemist. I remember that in ~1938, just before World War II, my mother, living in a marshy area just outside Amsterdam, got malaria. Thanks to DDT malaria was eradicated from the Netherlands shortly after World War II.

245 Priester (1998: 62); De Meere (1982) – The malaria outbreak in Amsterdam, 1846, caused many excess deaths in the severe winter 1846/7 due to under-nourishment. Also Jansen & de Meere (1982).

246 We do not know whether these peaks are statistically significant.

247 A sketch of the abject situation in the Jewish quarter of Amsterdam was given by the Allgemeine Zeitung #183 dd 02-07-1846: 1460 (Augsburg Edition). For disease and mortality rates see Jansen & de Meere (1982: 186ff).

248 The Netherlands – Hofstee (1978: 210); Terlouw (1971: 288); ~6,000 deaths in Belgium – Lamberty (1949: 136); ~146,000 deaths in France – Dupâquier (1988: 293); Ireland – Powell (1998: 1106).

249 Belgium – André & Pereira-Roque (1974); Flanders – Lamberty (1949: 136); Ó Gráda (1989: 60); West Flanders – Hofstee (1978: 199).

250 Estimate for Prussia based on tables by Mohorst (1977) and Kraus (1980). Data for East Prussia from Wehler (1987: 653), for Upper Silesia from Lutz (1985: 444). Abel (1974: 388) comparing 1844 and 1847 stated that deaths in Prussia increased from 403 to 512 thousands, implying 109,000 excess deaths. Diseases and figures vary. Wehler (1987: 653) speaks of typhus with 30,000 deaths in Upper Silesia, and 50,000 (excess?) deaths in total.

251 More precisely: the pre-industrial chronic poverty of large numbers of people in a region and/or period. The term seems to be typical for the early and mid 19th century.

252 The Netherlands – Brugmans (1929: 73); Flanders – Lamberty (1949: 135); the Germanies – Lutz (1985: 84, 94).

253 Mr. Ph.J. Bachine, a respectable gentleman living in the city of Sluis, near the border with Flanders (Belgium), complained that up to 200 beggars a day knocked on his door in the early winter of 1845 (*ex* Kroes, 1987: 34).

254 De Joode (1981: 143); original source not known.

255 '... *la misère... à l'état endémique...*' (Agulhon *et al.*, 1976: 78).

[256] Most historians think that the economic crisis in France began as a rural crisis but Droz (1967: 267) inverted the argument. Agriculture was dragged into the crisis because landowners withdrew money from agriculture to invest in movables.

[256a] '... *des explosions de colère dans les masses rurales*.' ('... explosions of fury among the rural masses.' Agulhon *et al.*, 1976: 78).

[257] For a recipe see von Bujanovics (1847).

[258] Guin (1982: 116).

[259] See Goebel's (1980) insightful analysis of '*Les Paysans*'. This novel is situated in about 1844. The first part appeared in serial form in '*La Presse*', 1844. The complete text was published posthumously in 1855. The fictional but factual content and its moral background are precisely confirmed by historical evidence as presented by e.g. Vigreux (1998).

[260] Armengaud (1961: 179) '*Plus de 25,000 personnes touchent au moment de n'avoir plus d'aliments*' and at Rimont two thirds of the inhabitants '*manquent de tout: sans argent, sans pain, sans pommes de terre, sans travail, sans crédit enfin*'.

[261] Agulhon *et al.* (1967: 399), Newman & Simpson (1987: 897), Vigreux (1998: 227/238).

[262] Bergman (1967: 405) shows a map of Dutch towns with riots in 1845/7.

[263] Vigreux (1998: 233) commented that the straw-thatched cottages could have caught fire spontaneously, or by imprudence, during the excessively hot and dry summer of 1846.

[264] Jardin & Tudesq (1973: 236); Vigreux 1998: 232/3).

[265] Berlin, April 21-23, 1847, had its potato revolution (G: *Kartoffelrevolution*) with pilfering of food shops; the revolutionaries had no political aims (Droz, 1957: 109).

[266] Hoekstra (1879); Hoogerhuis (2003: 31); Priester (1998: 60); Terlouw (1971: 290). In the period 1845-1847 some 1100 Dutch families sailed for America, among them many Frisian farmers (Wumkes, 1934: 222). Probably, the number is too low.

[267] Bergman (1967); Ising (1892); Terlouw (1971).

[268] The rule was confirmed by its exceptions, among them the parsons Heldring (1845) and Hooijer (1847), who earnestly tried to organise relief work.

[269] E.g. Abel (1974: 379); Armengaud (1961: 180); Jardin & Tudesq (1973: 236) – Toulouse and Bordeaux.

[270] The very Irish gentleman Austin Bourke (1993: 178) stated 'It should be said right away that there is not a title of evidence during these years which points to deliberate malintent towards the Irish people on the part of the British government or its representatives in Ireland. There is, on the contrary, ample proof, in all their deliberations and acts, of a sincere desire, within the then accepted ideas of the limits of legitimate government action, to alleviate the crisis brought on by the potato failure.'

[271] De Bosch Kemper (1882: 68). Note that this action is meant to be the opposite of 'hoarding', merchants buying grain and keeping it until the price had gone up, as in France, 1846 (Agulhon *et al.*, 1976: 141).

[272] Hoekstra (1879) in 1846: *Men heeft hier te Irnsum voor de gemeene man het brood afgeslagen tot op 5 stuivers, waarvoor de uitgaven uit een kollect bij de notabelste ingezetenen gevonden worden.*

[273] Newspaper clip 31 May, 1847 (Wumkes, 1934: 219).

[274] Virchow (1848: 39/41). Breslau is present Wrocław in S Poland. *Das Breslauer Comité, welches erst aus ganz Deutschand Geld zusammenbetteln musste, war eher auf dem Platz, als die Regierung!*

[275] Lette, XX (1847: 28) *'Über die bäuerlichen Creditverhältnisse der Provinz'*. (On the farmers'credit situation in the province).

[276] Seelman-Eggebert (1928: 62) *Was nun die Geschichte dieser Vereine betrifft, so ist deren Geburtsstätte der untere Westerwald, in der Preussischen Rheinprovinz, die eigentliche ursprüngliche Zeit der Entstehung das Notjahr 1847.*

[277] Terlouw (1971: 291) ascribes the financial crisis in London to grain speculation; Woodham-Smith (1962: 408) mentions the financial crisis of 1847 without comment. See also Reader (2008: 184 ff).

[278] Paris – Furet (1988: 372); Newman & Simpson (1987: 898). New York, London, Frankfurt – Wehler (1987: 651); London – Ó Gráda (1989: 46); Vienna – Rumpler (1997: 276). Bickel (1974: 123) stated that this financial crisis was the first that originated in the USA but Wyckoff (1972: 223) wrote that the crisis at the New York Stock Exchange was imported from Europe.

[279] ICT = Interactive Computer Technology.

[280] The German daily *Allgemeine Zeitung*, usually avoiding detail, reported on railway shares in Europe every day of July and August, 1848.

[281] Rabl (1906: 149), letter dated 18 May, 1848, that is 2 months after the March Revolution in Berlin: '..., *täglich werden neue Massen von Handwerkern brodlos, die Fabriken stehen eine nach der andere still u. wir Alle werden in unserer Nährung gestört'*.

[282] Ackersdijck *et al.* (1852: 324): Paris industry produced a value of 1.46 billion French Francs in 1847 but only 0.67 in 1848, a decrease of over 50 per cent. Some 186,000 workers out of 342,000 were without work, again over 50 per cent.

[283] Houssel (1976: 236), Furet (1988: 374), Jardin & Tudesq (1973: 238).

[284] Lutz (1985: 244); Wehler (1987: 641); many others.

[285] Virchow (1848: 13) ... ,*das Kind an der Mutterbrust wurde schon mit Schnapps gefüttert'*.

[286] Dutch citizens subscribed to the loan, not without some moral pressure.

[287] Bergman (1967: 418), Terlouw (1971: 272).

Notes

288 See e.g. Sen (1981); Drèze & Sen (1989).

289 Leonhard (2006: 11), e.g. Baden, Bavaria, Württemberg, but not Prussia and Austria.

290 The 19th century liberalism bears little relationship with the current idea of a 'liberal' in the USA around 2000, where it is nearly a term of abuse. The Constitution of the USA (1787) typically resounds a 'liberal' tone, liberal in its classical meaning.

291 This gathering of March, 24th, 1848, was later described as emancipatory and non-revolutionary though it ended inevitably with some window smashing and pilfering (Brugmans, 1929: 188). Nonetheless it impressed the King, an emotional person. According to Fuchs (1970: 58/9) a foreign communist agitator manoeuvred in the background.

292 De Beus (2006: 86): 'It entailed sovereignty of the Dutch state, constitutional monarchy, bicameral parliament, ministerial responsibility and unity of cabinet policy, separation of state and church ..., basic rights and liberties for all citizens,... public utility monopolies, assistance of the poor, education, and arts and sciences'.

293 The Second French Republic under Cavaignac's new 'Constitution of the Republic' held until 2 December, 1852, when Bonaparte became Emperor of France, ruling as an absolute monarch with a lame Parliament.

294 Droz (1957: 111) *'avait ébranlé l'autorité et le crédit de l'État'*.

295 Mitchel with his 'Young Ireland Party' and a group of peasants attacked the police station of Tipperary. Mitchel was caught within a week (Woodward, 1938: 3342).

296 Gotthelf, a pious protestant, warned against the 'liberals'. Gotthelf (meaning 'so help me God') was the pseudonym of Albert Bitzius, a prolific writer with a keen eye for actuality.

297 Mommsen (1998): *Die Revolution von 1848 war nicht das Produkt zielbewusster revolutionärer Aktionen: sie war eher eine Implosion der überkommenen Statsordnung auf dem Europäischen Kontinent, deren Legitimitätsgeltung schlagartig eingebrochen war und für deren Erhaltung zu Kämpfen mit einem Male als aussichtslos empfunden wurde.'*

298 Wehler (1987: 659) *Einen direkten Nexus zwischen Hungerkrise, Getreidetumulten and Kartoffelrevolten einerseits, der Revolution von 1848 andererseits zu unterstellen, hiesse, einer simplifizierenden Verkürzung Komplizierteren Zusammenhänge zu erliegen* and he added ... *der Krise von 1845 bis 1848, das Präludium zur 48er Revolution in ganz Europa.*

299 Pirenne (1932: 128) *La crise économique se complique bientôt d'une crise alimentaire.*

300 Lutz (1985: 244) *Die wirtschaftliche Strukturkrise der vierziger Jahre, die durch Bevölkerungsdruck, Massenarbeitslosigkeit und Pauperismus gekennzeichnet war, wurde 1846/7 durch eine akute Krise verschärft. Die Kartoffelkrankheit und Missernten führten zu einer Hungersnot, die sich in Epidemieen und in einem Rückgang der Geburtenrate äusserte.* (ibid.: 245): *So erwuchsen aus Hungersnot und Hungerepidemie oppositionelle Kritik und öffentliche Bewusstseinsbildung,* and (ibid.: 245) *Ein unmittelbarer Zusammenhang zwischen den Hungerunruhen 1847 und der Revolution wurde auch von nichtsozialistischen Zeitgenossen angenommen.*

301 Sneller (1943: 436). ... *het gaf tevens den stoot mee tot de vreedzame omwenteling in het landsbestuur in het jaar 1848* ....

302 Le Roy Ladurie (2004: 623): *Sévit enfin, s'agissant toujours d'événementiel et de variabilité, la longue et chaude sécheresse du printemps et de l'été 1846, laquelle, avec la maladie des pommes de terre, notamment irlandaises, porte certaines responsabilités via la dépression économique de 1847 née de cette mauvaise récolte; elle implique effectivement une espèce de culpabilité climatique, fût-elle partielle, vis-à-vis du déclenchement ultérieur de révolutions de France et puis d'Europe occidentale et centrale, à partir de février 1848, dans l'ambiance d'une crise de l'économie qui procède en effet de la difficile année-post-récolte 1846-1847; dans l'ambiance aussi, corrélative, d'un certain mécontentement des peuples divers.*

303 Ising (1892: 47) '... *landbouwer, die voor het mislukken van zijn aardappelen in de hooge prijzen zijner granen meer dan vergoeding vindt* ...'

304 Rüter (1950: 524): *Het afgeloopene jaar is voor den landbouw in dit gewest, over het geheel, zeer voordeelig geweest, niet zoozeer door een rijken en overvloedigen oogst, als wel door de hooge prijzen van alle voortbrengselen.*

305 The weight of the evidence presented in this paper is variable. Population statistics were sufficiently accurate for the present purpose, migration data excepted. Yield statistics were not readily available except for the excellent data from Belgium, 1846, and the more approximative Prussian data. Most yield data are from quotations by historians who did not mention their sources. Data on the late blight epidemic, 1845, from the Netherlands, Belgium and Prussia are considered fairly reliable though not very accurate; independent and contemporaneous confirmation was found in the *Staatscourant* (Appendix 3.1). The 1846 data from the Netherlands, however, are vague. For the Netherlands contemporaneous and independent confirmation was found in the newspaper *Nieuwe Rotterdamsche Courant* with its daily updates. With increasing distance from a nuclear area, roughly the Netherlands, Belgium and W Prussia, the information became less accessible, vaguer, lighter. Nonetheless, I believe to have given a fair and consistent picture of potato late blight and its consequences in Continental Europe.

306 Rotting during transport was mentioned several times. In Bremen a shipload of potatoes was thrown in the river because they were rotting (Fürnrohr, 1845: 935). Van Rossum (1845) received about 25 October, 1845, 1500 'mud' (= ~1000 hl) of good quality potatoes from Stettin (= Gdańsk) but some tubers were rotten and several others infected.

Chapter 4

307 Unpublished biographical note by my grandfather's eldest son, then 26 years old, a person blessed with an absolute memory. So far I did not find a confirmation in Amsterdam police reports. Dr Nicolaas Marinus Josephus Jitta, ophthalmologist and 'hygienist' (= epidemiologist), was chosen a member of the Amsterdam City Council, 1899, for the Liberal Union. In 1905 he became the Alderman for Public Health and Market Trade. An administrator rather than a politician, he relinquished late in 1917, feeling deeply hurt by the violent attacks of the opposition in the City Council after the riots.

308 Amsterdam Municipal Archive, minutes of the City Council meetings. Gemeenteblad van Amsterdam 1917 Vol. I-2: 859-868 (Mayor's answers to interpellations in Council Session), Vol. II-2 (Public discussion in Council, 11 July 1917)

[309] World War I began 1 August 1914, when Germany declared war to Russia, and ended 11 November 1918 with the armistice signed at Compiègne in France. The peace treaty was signed at Versailles (France) on 28 June, 1919. From an agricultural point of view the harvest seasons 1915, 1916, 1917 and 1918 are of interest.

[310] Potatoes had been rationed temporarily.

[311] The Dutch farmers' weekly *De Veldbode* (= The Field Messenger) did not discuss potato late blight in 1915, 1916, or 1917. The profitability of Bordeaux mixture was mentioned casually in an unrelated paper (Vol. 14 – 1916: 112). *Overijsselsch Landbouwblad* (Agricultural Journal of Overijssel), 1-3 (1916-1918).

[312] Annual Reports 1903-1930 (1851-1960).

[313] Annual Report on 1921: 26.

[314] In a more formal treatment we should relate the variables d, y_o and y_e to the specific year t under consideration: $d_t = y_{o(t)} - y_{e(t)}$, and $\delta_t = d_t / S_x$. The index t has been omitted.

[315] H.M. Quanjer, pharmacist by training, assistant of Professor J. Ritzema Bos in Amsterdam and Wageningen, later Professor of Phytopathology at the Wageningen Agricultural College, founder of the Laboratory for Mycology and Potato Research at Wageningen, the only 'European Correspondent' ever of the American journal 'Phytopathology'.

[316] This statement, not infrequent in Dutch and German literature of the time, is contrary to recent experience. Two explanations, not mutually exclusive, are proposed. (1) The late blight appeared late, when the early potatoes had already completed their growth, as suggested repeatedly. (2) In recent times the late varieties have 'field resistance', a characteristic selected for after World War II. I surmise that the late varieties around 1900 were, in modern eyes, highly susceptible. During World War I a cultivar such as 'Bintje' was coming into favour and its relative susceptibility apparently was not an obstacle; it fell out of grace around 2000 when it was considered far too susceptible.

[317] Quanjer (1911, 1912, 1914); Ritzema Bos (1919).

[318] See for illustration Bieleman & Van Otterloo (2000: 212).

[319] See Table 7 in Van der Poel (1967: 213). Spraying machine dated 1913 in Van der Poel (1967, Figure 82: 212).

[320] Annual Report on 1917 (XVIII): little spraying because of dearth of $CuSO_4$.

[321] Van Bavegem (1782); Uilkens (1853); Vanderplank (1949).

[322] Anonymous (1910: XXI); Annual Report on 1913: XVI.

[323] Salaman (1949) wrote 1904. Otto Appel (1906a) introduced the name *'Blattrollkrankheit'* (= leaf roll disease) for the new potato disease of 1905, resembling the old 'curl' but sufficiently distinct to be given a separate name. In Denmark Rostrup (1906) called 1905 a typical insect-year (Danish: *Insektaar*). In the Netherlands the summer index of 1905 was 60 (Figure 4.1A). Both data suggest a fine summer with little rain, that might have been favourable to aphids and hence to dispersal of viruses.

[324] Combine harvesters were not yet in use.

[325] *Septoria tritici* (*Mycosphaerella graminicola*) and *Septoria nodorum* (*Leptosphaeria* (*Stagonospora*) *nodorum*), not to mention various *Helminthosporium* and *Fusarium* diseases. See historical overview on the Septorias by Zadoks (2003).

[326] The Annual Reports sometimes mention other sources of information providing slightly different figures.

[327] I have not found a mechanistic or physiological explanation for the effect. Possibly the good quality of seed potatoes, fully ripened and little infested by pathogens, is part of the explanation.

[328] At the time the potato harvest was not yet mechanised.

[329] Kielmansegg (1980: 177); Herzfeld (1968: 184).

[330] Herzfeld (1968: 185); Siney (1957: 36).

[331] 'Annual Report' on 1914: VII, LXXIII.

[332] 'Annual Report' on 1914: LXXIV ff, 1915: LXVIII ff, 1916: LXVII ff, 1917: LXVI, 1918: LIX. Farmers were invited to make fertiliser claims, and these were checked by the municipal authorities. Fertilisers were distributed in proportion to the approved claims.

[333] The Annual Reports of 1914 to 1918 contained short paragraphs on 'Agriculture and war' (D: *De landbouw en de oorlog*), that emphasised the transition from open export-oriented to closed self-supporting production. Rapid rises in production costs could not always be compensated by rises in product prices. The reports expressed great concern about the supply of fertilisers since most had to imported. However, yield figures do not convincingly indicate production losses due to lack of fertilisers. (1914: VII, 1915: VII, 1916: VII, 1917: VII, 1918: V).

[334] Annual Report on 1918: LXXVI. Practically all exports had been prohibited.

[335] If we assume, as a first guess, that starch potatoes at the time contained about 20% starch, the potato flour exports were roughly equivalent to the fresh potato exports. Other potato products can be neglected here.

[336] As an Alderman my grandfather offered cheap rice, of which there was ample stock, to the protesting Amsterdam women but they refused the offer indignantly. One woman worded the reason 'If I put this [= rice] before my husband, I'll get hell' (Anonymous, 1917: 1460). Refusal to change the diet in times of scarcity is of all times and all countries, and may lead to a misplaced famine alarm.

[337] Typical is the dejected comment by the Mayor of Amsterdam in the city Council Meeting after the 1917 food riots: 'We need the cooperation of England and Germany. It seems necessary to buy that cooperation. To buy it with potatoes that we would prefer to keep ourselves.' (*Wij hebben de medewerking van Engeland en Duitsland nodig. De medewerking schijnt gekocht te moeten worden. Gekocht met aardappelen, die wij liever zelven zouden houden*) Gemeenteblad van Amsterdam 1917 Vol II-2: 1479.

[338] www.library.uu.nl/wesp/populstat/Europe/netherlc.htm. Accessed April, 2006.

Notes

[339] Opinions vary about the health effects of the scarcity (Bieleman & Van Otterloo, 2000: 256 ff), from pernicious to just tolerable.

[340] When the Germans bombed Antwerp, October 1914, about one million Belgians fled to the Netherlands. Some 30,000 arrived by train in Amsterdam (Kruizinga, 2002: 120). In the course of 1914 and early 1915 some 900,000 Belgians repatriated (Pirenne, 1928: 278).

[341] In June, 1916, the scarcity of potatoes was threatening and what was available had a poor quality. Demonstrations occurred in many Dutch cities.

[342] Among which, notably, Rotterdam, 1917 (Bieleman & Van Otterloo, 2000: 256 ff).

[343] Potato riot (D: *Aardappeloproer*). The spring of 1917 saw many protests again. The week following 28 June, 1917, riots, lootings and strikes occurred in Amsterdam and several other places (Huijboom, 1992; Kruizinga, 2002: 15). For Amsterdam, the *Gemeenteblad* I-2: 868 quoted 9 casualties, 114 wounded civilians, and looting of 8 shops, 5 warehouses, 1 hand-cart(!), 2 wagons, 3 railway vans, and 3 barges (Fuchs, 1970: 121, gives slightly different figures).
The government wanted to export potatoes to remain at good terms with England as well as with Germany. It was said that two wagons loaded with potatoes for exportation, stationed in the Amsterdam railway yard 'De Rietlanden', were looted by a crowd of hundreds of angry women. The official version, however, reads that these potatoes were being transferred to a barge with the grocery market as its destination, that the women exacted a sale and that the enforced sale – not without the consent of the owners of the potatoes - took place under the supervision of the police (Anonymous, 1917 II-2: 1297/8). Small quantities (3 kg per person) were sold at the official price. Women, who did not carry their purse with them, obtained the potatoes on credit (Anonymous, 1917 II-2: 1469).

[344] According to modern standards the enquiry was not representative; it was serious but possible prejudiced; its emotional impact was considerable.

[345] Annual Report on 1916: VIII.

[346] Kielmansegg (1980: 178) and various other German authors. Sneller (1943: 24).

[347] This paper did not consider grain in storage before the outbreak of the war nor grain imported in the early war years.

[348] The rust was probably yellow stripe rust (*Puccinia striiformis*); a yellow rust epidemic developed in S Germany, Bavaria (Hiltner, 1905). Yellow rust was prominent in Denmark on wheat, barley and rye (Rostrup, 1905: 355).

[349] In the period June through August there were large precipitation differences between N (< 200 mm) and S (> 300 mm; Annual Report over 1905: IX), visible in potato yields (*ibid.*: XVI). See Table 4.5.

[350] In N Germany 1905 was a typical blight year; simultaneously an epidemic of the 'curl' developed with 'leaf roll' as the prominent symptom (Appel, 1906: 122). No mention was found of curl, 1905, in the Netherlands.

[351] According to De Bokx (1982) 1906 was a typical curl year. Present author's comment: Seed potatoes could have been infected during the relatively fine (and possibly aphid-rich) summer of 1905.

[352] Groningen wheat area at least 25 % below mean of 1906 and 1908.

[353] Annual Report on 1908 (XIV) mentioned ´threatening curl in 1907´.

[354] Rain damage to wheat in July (Groningen) and in August (Zeeland).

[355] According to De Bokx (1982) 1908 was a typical curl year again.

[356] In De Bilt temperatures down to -16 °C, in Groningen some 20% of winter wheat frozen and ploughed under (locally up to 50%), but in Zeeland good yields though poor quality.

[357] In Brabant rye damaged by corn thrips (possibly *Limothrips cerealium*), in Limburg increase in stem nematode (*Ditylenchys dipsaci*) damage.

[358] According to De Bokx 1909 was another typical curl year.

[359] Comment not confirmed by Figure 4.3c. I have no explanation of the discrepancy.

[360] Probably due to the fungus *Gäumannomyces graminis*.

[361] Good effect of Bordeaux mixture. In Zeeland the 10% of fields that had been sprayed produced good yield and quality (Annual Report on 1913: XVI-XVII).

[362] The rust was most probably yellow rust (*Puccinia striiformis*). It affected mainly new wheat varieties from Svalöff (Sweden) and spring wheat Japhet. In 1914 yellow rust was epidemic from the Netherlands to far into present Poland, Austria and S Germany (various sources in Zeitschrift für Pflanzenkrankheiten, 1914-1918). Rust on barley might have been *P. striiformis*. This rust and *P. hordei* respond in nearly the same way to environmental conditions.

[363] Probably the brown rust (*Puccinia recondita*); it was severe in some areas of Germany.

[364] Probably barley dwarf rust (*Puccinia hordei*).

[365] Mainly black grass (slender foxtail, *Alopecurus myosuroides*, D: *wintergras, duist*).

[366] Verticillium wilt probably due to *Verticillium albo-atrum*. The Annual Report on 1918 mentioned an increase in potato area in reaction to higher prices (XVI) and an improved procurement of seed potatoes as field inspection services extended and farmers strived for change of seed potatoes (XVII).

[367] Disastrous late blight epidemic in Germany (Schick & Klinkowski, 1962: 1170).

Chapter 5

[368] These authors probably based their judgment on an explicit statement by Löhr vom Wachendorf (1954: 121), a statement not incorrect but exaggerated and too one-sided.

[369] The armistice was signed on 11 November 1918 in a railway carriage, in the forest of Compiègne, France. The blockade of Germany was continued except for food. The Peace Treaty was signed 28 June 1919 in Versailles, France.

[370] World War I affected all aspects of civilian life by propaganda, psychological warfare, food rationing, female and child labour, bombing of open cities, starving the enemy, and so on. It affected numerous countries and spread over Europe, Africa, Asia (Near and Far East), the Atlantic Ocan, and beyond.

[371] Mommsen (2004: 92/3).

[372] According to various authors, e.g. Hirschfeld *et al.* (2003: 461, 565); Mommsen (1995: 682).

[373] Flemming (1978: 82). Two million out of 3.4 million able-bodied men were drafted.

[374] 'Wet planting' (G: *nasze Bestellung*; Schander, 1917: 162). No explanation of this comment was found. Supposedly the small farmers were in a hurry to plant in view of the lack of labour.

[375] Weediness was stressed by Aereboe (1927: 38, 94). See also Illustrierte Landwirtschaftliche Zeitung 37#4 (1917) January 13th - Weeding is light work, can be done by school youth. Deutsche Landwirtschaftliche Presse 44 (1917: 43, 235) – Weeding by school kids was obligatory in Prussia as of 3 February, 1917.

[376] Schander (1917: 162); Flemming (1978: 86), overcropping, premature exhausting (G: *Raubbau*) of soil and people.

[377] E.g. Herzfeld (1968: 180); Flemming (1978: 96) 'As the dogma of the war's short duration had prevented prospective activities for an economic mobilisation in peace-time'.

[378] The Haber-Bosch procedure produced ammonia from H_2 and N_2 in the air applying a large input of energy (from coal). Ammmonia was oxydised to nitrate. Industrial application of the procedure began in 1913.

[379] In my experience rusts, mildew, and ripening diseases of cereals often have a reduced impact on yield when nutrient supply, especially of N, is poor. No comments to this effect were found.

[380] See also §4.4.1. D: *Degeneratie, ontaarding, verval, veroudering*. E: Degeneration, deterioration, running out. F: *déchéance, dégéneration*. G: *Abbau, Entartung, Verfall, Herabzüchtung*. The phenomenon was well known but attitudes differed. The German Hoppe was the first to describe the disease, in 1747 (Salaman, 1949). The German Kühn (1858) thought it nonsense, being only of local occurrence. Quanjer (1921a) identified several non-filterable agents that could be reproduced in pure culture (*in planta*). French (e.g. Blanchard & Perret, 1917) and German researchers thought of environmental conditions primarily, though they did not negate 'vertical transmission'. German resistance to the idea of viruses as causal agents re-appeared in the review by Morstatt (1925: 44), who quoted a colleague 'The Dutch concept of degeneration cannot be considered as generally correct' (G: *Die Holländische Auffassung des Abbaues kann nicht als allgemein richtig gesehen worden*). Morstatt mentioned the word 'virus' rarely (*ibid.*: 50, 64) and not at all in his conclusions where 'local conditions' (G: *Standortsbedingungen*) is the one-but-last word. Several potato viruses are transmitted by aphids which thrive during periods of dry weather.

[381] The epidemic spread to the Netherlands (De Bokx, 1982), Denmark (Lind, 1916a), and other countries.

[382] Denial of Quanjer views (e.g. Schander & Tiesenhausen, 1914; Esmarch, 1919; Neger, 1919) was common in Germany. Only A. Mayer quoted Quanjer in a positive sense (Fülings Landwirtschaftliche Zeitung 65:

474-478). O. Appel acknowledged the idea of an infectious contagium (Fühlings Landwirtschaftliche Zeitung 67, 1918: 85).

[383] Blanchard & Perret (1917).

[384] Rozendaal (1949: 105), Van der Zaag (1999: 96).

[385] Remy (1916); Remy was stationed in Bonn, on the river Rhine, in W Germany.

[386] Compare seed potatoes grown in Scotland and exported to England, and those grown in Friesland and exported to other parts of the Netherlands (Van der Zaag, 1999).

[387] German rail transportation was excellent. During the war it was made subservient to military purposes. It became a total mess only during the winter 1916-1917 when, unfortunately, transportation by water had become impossible because all waterways had been frozen (Herzfeld, 1968). In addition, recent regulations ruled against seed potato exchange. Remy (1916: 818) complained about difficulties in shipping seed potatoes from E to W Germany; see also Schander (1917: 150).

[388] Losses as quoted by Schander are not too bad. Wennink (1918) found that loss due to leaf-roll could be up to 90 per cent.

[389] Quanjer *et al.* (1921a: 13), Schander (1918: 219).

[390] Schander (1918: 220); in the Netherlands good results were obtained by spraying Bordeaux mixture (Quanjer, 1912, 1914; Ritzema Bos, 1919), but in the war years copper was scarce and costly. In Denmark, two applications of Bordeaux mixture could prolong crop life by one month, thus boosting crop productivity (Lind, 1916b).

[391] References to lack of Bordeaux mixture were found in Dutch reports (Annual Report over 1917, and 1918: XVIII; Chapter 4).

[392] Schick & Klinkowski (1962) mentioned this interaction between potato late blight and the potato viruses X and Y without details. A loose, contemporaneous comment was made by Kuhn (1918), 'degenerated planting material was more susceptible to disease'.

[393] Anonymous (1919: 47/9). The term 'blackleg' is indicative of a symptom, blackening of the stem base of the potato. The most common cause is a bacterial disease, due to *Erwinia atroseptica*. This disease, transmitted a.o. by seed tubers, is favoured by continuous moist weather and by wet soils. When the green plant is affected yield decreases, when the tubers are affected they may rot in storage (Appel, 1928: 222ff).

[394] Mitteilungen aus der Biologischen Reichsanstalt für Land- und Forstwirtschaft 16 (1916: 59).

[395] On the 1914 yellow rust epidemic: *the Netherlands*: Annual Report on 1915, mentioning severe yellow stripe rust infection in Groningen, in the NE, especially on varieties of the Swedish breeding station Svalöf. *Germany*: Anonymous (1916), central station received many samples; Müller & Molz (1917), much yellow rust on wheat in Saxony, heavy dews; von Wahl & Müller (1915), on wheat in Baden; Zimmermann (1916a), explicit epidemic on wheat in Mecklenburg; Present *Poland*: Oberstein (1914), Silesia, mainly on wheat, also on rye; Schander & Krause (1917), yellow stripe rust on wheat with unusual severity (G:

ungewöhnlicher Heftigkeit) in Posen (Poznań) and W Prussia near Bromberg (= Bydgoszcz) on the river Weichsel (= Visła), rye also infected; *Austria*: Anonymous (1914), on wheat and rye; Kornauth (1915), on wheat and rye.

[396] Vole attacks in 1914: Oberstein (1914) – Silesia, severe attack by voles mainly in clover; Schander (1915, 1917) – same; Zimmermann (1916) – really exceptional were damages in wheat by voles, locally crops were beyond hope.

[397] In Chapter 3 the name 'Prussia' referred to the Kingdom of Prussia and/or to its E parts; in Chapter 5 it refers to the NE provinces of the German Empire.

[398] Schander & Krause (1915: 215), Zimmermann (1916b: 321).

[399] Control of voles in the field could be obtained by frequent tillage, at home in the granaries by immediate threshing and strict hygiene. Control by means of CS, poisoned baits, and mouse traps was also applied; these were labour-intensive methods (Schander & Krause, 1915: 216; Schander & Krause, 1917: 129).

[400] Carefoot & Sprott (1967: 87); Löhr vom Wachendorf (1954: 113).

[401] Several potato viruses are transmitted by aphids which thrive during periods of dry weather.

[402] Schander (1917).

[403] Löhr vom Wachendorf (1954: 114), spring 1916 stench vacation (G: *Stinkeferien*). Löhr's book was written in the style of a novel, but his facts were usually correct. The tragedy revealed by Löhr was quoted by Carefoot & Sprott (1967: 87). Löhr apparently was a soldier at the time (*ibid.*: 116) and an eye witness of this shameful episode with '*fast verbrecherische Verplempung*'. We do not know whether these storage losses were local, and Löhr's personal experience only, or general. I have not met with an independent confirmation of Löhr's data, not even in the knowledgeable and critical treatise by Aereboe (1927), except for an allusion by Schander (1918: 224). Is it by chance that Schander & Krause (1917: 130) reported in 1916 on experiments in potato storage during the winter 1913-14? Storage in 'pits' (G: *Mieten*) was better than in warmer storage rooms.

[404] Schander (1918: 224) '*Im Jahre 1916 hätten tausenden von Zentnern Kartoffeln erhalten und für die menschliche Ernährung nutzbar gemacht werden können, wenn die Trocknung rechtzeitig eingesetzt hätte.*'

[405] It may not be accidental that G. Schneider published three papers on potato storage in 1918.

[406] At the time, *Fusarium* research was in its infancy. The data are not very clear. Snow mold was well known, as was red discolouration of wheat and rye grains.

[407] Mommsen (1995: 692). Honcamp (1918: 436), who liked exaggeration, spoke of a completely failed potato crop (G: *völlige Kartoffelmissernte*). In England the potato harvest was disastrous too (Salaman, 1970: 574). In the Netherlands potato yields were low because of curl and late blight (Chapter 4).

[408] The enemy country, England, had a good potato harvest in 1918, mainly due to a great extension of the potato hectareage (Salaman, 1970: 576), as contrasted to Germany with a shrinking hectareage.

[409] The quality of seed potatoes is determined by genetic characteristics, a variety of environmental factors during the growing season and the storage period, tuber handling, and pests and diseases. At the time some empirical knowledge was available but the publications suggest much guesswork in the explanation of seed quality characteristics. More systematic knowledge was acquired during the second half of the 20th century (Struik & Wiersema, 1999). Reverse projection of present knowledge on forgotten varieties and old practices is risky and has not been attempted.

[410] R. Schander was stationed in E Germany, in Bromberg, Pommerania (now Bydgoszcz in Poland).

[411] The 'export limitation' (G: *Ausfuhrbeschränkung*; Schander, 1917: 150) apparently was a regional Prussian affair.

[412] Seed potatoes were taken from healthy fields. Apparently, field inspection with elimination of diseased plants was not yet practiced. Pedigree selection was known but still unusual.

[413] In the Netherlands the hot and dry year 1959 was a typical aphid year leading to a severe outbreak of potato virus Y^N; similarly, the very dry and warm year 1976 saw an explosion of potato virus Y^N (de Bokx, 1982: 24). The text suggests that in the case of virus Y^N the disease exploded in the dry year itself. In Germany, however, the effect of wide-spread virus infection (here not Y^N) during 1915 became manifest in the following year, 1916.

[414] Black leg was and is attributed primarily to a bacterial disease caused by *Erwinia carotovora* subsp. *atroseptica* (Van Hall) Dye, which flourishes during periods of prolonged wet and warm weather.

[415] Mitteilungen aus der Biologischen Reichsanstalt für Land- und Forstwirtschaft 17 (1919: 17, blackleg: 48).

[416] Mitteilungen aus den Kaiserlichen Anstalt für Land- und Forstwirtschaft 17 (1919: 17), the harvest 1916 produced only nut-sized potatoes.

[417] Leaf roll – a 'degeneration' disease due to a potato virus. An early epidemic occurred in 1905 (Appel, 1906). The disease is transmitted by aphids. Ritter (1917) said that 1916 was an insect-poor year. In the field the symptoms of leaf roll can easily be confounded with symptoms caused by the fungus *Rhizoctonia solani* (Appel, 1917) and the fungus *Verticillium alboatrum* (Quanjer, 1921b).

[418] In view of the lengthy explanations in the paper by Schander we cannot avoid the suspicion that his conclusions, published in 1917, were politically coloured. The complete absence of any reference to potato spraying feeds this impression.

[419] Mitteilungen aus den Biologischen Reichsanstalt für Land- und Forstwirtschaft 17 (1919: 47).

[420] On the 1916 yellow rust epidemic on wheat and rye: *Austria*: Fruwirth (1916), empty spikelets and even empty ears; Kornauth (1918), yellow rust very severe in Moravia and N Lower Austria (G: *herrschte sehr arg*), also in Bohemia; *Germany*: Halle (Upper Saxonia) - Müller & Molz (1917), Holdefleisz (1917); Samples of wheat with yellow rust, brown rust, and/or *Gäumannomyces graminis* exceptionally numerous (Anonymous, 1919: 47) Report on 1916; early summer was cold, much *P. striiformis*, *P. recondita* and *G. graminis*); Schander & Krause (1917), Prussia (Posnań, present Poland), yellow rust on wheat in early May already; Mitteilungen aus der Biologischen Reichsanstalt für Land- und Forstwirtschaft 17 (1919: 47); for yellow rust epidemics see Hogg *et al.* (1969); Zadoks (1961); Chapter 2.

421 Herzfeld (1968: 265); Mommsen (1995: 692). In the Netherlands too the coal distribution got stuck by the freezing of the canals (Smit, 1973: 10).

422 Schander (1918: 104); Deutsche Landwirtschaftliche Presse 44 (1917: 407).

423 Honcamp (1918: 436) *'Dürre des Jahres 1917 ... ungenügende Kornernte'.*

424 Mitteilungen aus der Biologischen Reichsanstalt für Land- und Forstwirtschaft 17 (1919: 59).

425 Schander (1918: 213 – leaf roll, 216/8 – late blight, 221 – loose skin, 217 – other diseases).

426 Honcamp (1918: 437), on Germany, 'splendid potato harvest' (G: *Glänzenden Kartoffelernte des Jahres 1917*). Salaman (1970: 576) on England, potato harvest 1917 very good. In the Netherlands the potato harvest was good (Chapter 4).

427 Deutsche Landwirtschaftliche Presse 44 (1917: 263). In the Netherlands winter killing of fall sown wheat was in the order of ten per cent (Chapter 4). In France, Paris region, over 65 per cent of the winter wheat was winter-killed (Comptes Rendues de l'Académie d'Agriculture de France, Séance du 25me d'Avril 1917: 427/8).

428 Honcamp (1918: 436), Kielmansegg (1980: 181).

429 Deutsche Landwirthschaftliche Presse 44 (1917: 697).

430 Vole attacks in 1917/8: Deutsche landwirtschaftliche Presse 44 (1917: 697); Schwarz (1920) large outbreak, 1918, in S Germany.

431 Hiltner, in Fühlings Landwirtschaftliche Zeitung (1918: 435).

432 Kornauth (1919), *Zabrus tenebrioides* Goeze (G: *Getreidelaufkäfer*) greatly damaged cereals, mainly rye, in Moravia, 1918, where several fields had to be ploughed.

433 Frey (1998), Siney (1957: 248), and several other authors.

434 Many authors, e.g. Flemming (1978: 88), Mommsen (1995: 693). The second economy amounted to at least 10 per cent of the national economy.

435 Flemming (1978: 78); Hirschfeld *et al.* (2003: 616); Mommsen (1995: 688); Mommsen (2004: 95, 133).

436 Honcamp (1918: 439), ... *ganz unverständlichen Wirtschafts- und Preispolitik.*

437 Aereboe (1927: 48) G: *völlige Kopflosigkeit.*

438 Honcamp (1918: 442) G: *hemmender behördlicher Masznahmen.*

439 Protest demonstrations in Berlin, 1915/6; January and February, 1917, in several cities (Mommsen, 1995: 692; 2004: 95); January, 1918 (Hirschfeld *et al.*, 2003: 565); Strikes in the *Ruhrgebiet*, the industrial

area along the river Ruhr, in early 1917 (Mommsen, 1995: 692) and climaxing in Leipzig, April, 1917 (Mommsen, 2004: 95).

[440] Grain from Romania was taken as booty, good to feed Germany during one month; as a joke it was mentioned that the British had already paid 90,000 tonnes (Deutsche Landwirtschaftliche Presse 44 (1917: 472).

[441] Swede = *Brassica napus* var. *napobrassica*, grown primarily as cattle fodder (Hirschfeld *et al.*, 2003: 461).

[442] Mommsen (1984: 362); The text quoted by Vocke (1984: 362), translated here, is nearly identical. The year of communication, 1917, may be mistaken. No distinction was made between hunger and hunger-associated disease.

[443] Remarque (1929). This bestseller was the first of a flush of German 'front novels'. It was banned and burned by the Nazis (Rüter, 1980).

[444] Aereboe (1927: 106) *'Richtig ist nur, dass die Schwierigkeiten der Ernährung der Menschen an der Front und im Lande schliesslich die Wiederstandskraft des deutschen Volkes gebrochen haben'*. See also Löhr vom Wachendorf (1954: 121), Kielmansegg (1980: 172).

[445] Also Flemming (1978: 88).

[446] Bumm (1928 Vol. I: 40, 41, 48, 57, 98).

[447] Hirschfeld *et al.* (2003: 565): as in Vienna, winter 1917/8.

[448] Several authors, e.g. Mommsen (2004: 125), who mentioned the decrease of *Leidensfähigkeit* among civilians, the 'ability to suffer'.

[449] A comparable story could have been written for Germany's partner in war, the Austrian-Hungarian double monarchy (e.g. F. Hilmer, 1916).

[450] Honcamp (1918: 41) *'... dass während des ganzen Krieges in landwirtschaflicher Beziehung der Wettergot nich auf unserer Seite gewesen ist, nicht einmal auf der neutralen.'*

Chapter 6

[451] Chrétien-Guillaume Lamoignon de Malesherbes, president of the 'Cour des Aides' (High Tax Court), in the 'remonstrance' to King Louis XV of France, 18 February 1771 (Badinter, 1978: 156).

[452] Le Roy Ladurie (1967: 621), Matossian (1989: 84).

[453] See Lefèbvre (1932), maps simplified by Matossian (1989: 85). Most but not all historians agree with the idea of a focal spread of the panic.

[454] No ergotism occurred in Germany where people had learned to avoid toxic grain. I did not find data on ergotism in the Netherlands.

[455] The possibility of ergot poisoning was also mentioned (independently?) by Le Couteur & Burreson (2003: 240): 'a bout of insanity in the peasant population with 'bad flour' as possible cause.'

[456] Tessier (1783: 103). '*Les habitans de Sologne, où la maladie gangreneuse a régné le plus souvent, & où l'ergot est plus abondant qu'ailleurs, ne vivent, pendant les trois premiers mois qui suivent la récolte, que de pain fait de seigle, en y comprenant le son*'.

[457] In Darwin's (1800) classification the 'external diseases' were due to external causes such as frost and heat. This Darwin was the grandfather of Charles Darwin.

[458] See e.g. Adanson (1763: 45); Bulliart (1791: Plate 111); Plenck (1794: 157)

[459] Tessier (1783: 25 – ergots per ear; 31 – season & locality; 35&80 – fallowing; 31 – reclamation; 71&78 – watering; 80 – experiment; 80 – ploughing; 28, 63, 83 – mixed cultivation, 28&83 – tillers; 81 – wetness; 69 – facts; 73 – insects; 78 – irrigation; 84 – explanation unnecessary.

[460] Tessier (1783): precautions p102; witnesses p103; dosages p103&107; ducks p107&110; hog p113; dog p122.

[461] The term 'noxious vapours' recalls the 'miasmatic' theory of disease, of malaria in special, current in the 18th century (and long after). The Solognese suffered malaria annually, mainly in autumn (Bouchard, 1972: 48). From September to May frequent ill-smelling fogs occurred, the stench due to tannery and rotting of hemp.

[462] Tessier (1773: 108 – health situation, 160 – abortion, 161 – nursing, 103&184 – rye consumption).

[463] Barger (1931: 80). German report not consulted.

[464] The episode is mentioned by Le Couteur & Burrreson, 2003: 240, who estimated the loss of troops at 20,000.

[465] Quotations from the Great Soviet Encyclopedia – 1978. New York, Macmillan. Vol. 19: 456.

[466] Graham (1929) is not explicit about the fate of the army near Astrakhan. He mentioned an army of well over 100,000 men in total, an army too large to be provided for (and far more than the Great Soviet Encyclopedia mentions).

[467] Conquest (1986: 117): On 27 December 1929 Stalin announced 'the liquidation of the kulaks'. The Ukrainian kulaks had already been reduced to small though independent farmers, with some three cows and one paid worker each.

[468] An extensive literature exists on the sufferings of the Ukrainian people due to the Stalin regime, e.g. Conquest (1984).

[469] Greenough (1982: 314): Birth rate in 1944 was 0.89 per cent versus 1.20 and more in normal years.

[470] In the next paragraph I suggest that the radiation deficiency of the 1942 season may have weakened the rice. It is known that weakened rice plants, apparently having lost their usual resistance, easily

succumb to brown spot (Klomp, 1977). If so, the host crop changed and not the fungus. Obviously, the environmental conditions were favourable to sporulation, dispersal, and infection.

[471] '... the rice plant shows no adaptability in its photosynthetic ability to weak irradiation.' (Matsuo *et al.*, 1995: 601).

[472] I did not see this point, radiation deficiency, reducing rice growth and yield without brown spot and making rice more susceptible to brown spot, mentioned in Indian literature so far.

[473] Sen-Gupta (1945): Yields of the 1942/3 season were poor but not far below average poor.

[474] For example, it is not clear whether the 1942 data incorporate the complete aman crop of 1942 that is harvested in the period December-January. Several yield and stock estimates have been said to be rough guesses made without on-the-spot visits.

[475] (Sen (1981: 52) and Brennan (1998: 545) mentioned the Midnapore cyclone.

[476] Uppal (1984). Tauger (2003), reaching back to Padmanabhan, follows an implausible reasoning.

[477] Incompetence of the Bengal Government was mentioned by Field Marshall Viscount Wavell, who, after installation as Viceroy of India, on 20 October 1943 visited Bengal already 26-28 October (Mansergh, 1973: #199).

[478] The newly formed country Pakistan consisted of two parts, West Pakistan (= today's Pakistan) and East Pakistan (= today's Bangladesh).

Printed in the United States
by Baker & Taylor Publisher Services